PRAISE

LIFE LE
FROM A
SURGEON

"Who knew a book about the human brain could be so enlightening and enjoyable at the same time? Science has come a long way, and Dr. Rahul Jandial is on the leading edge of some of the most promising and innovative breakthroughs in treating the brain. In *Life Lessons from a Brain Surgeon* Jandial strikes an elegant balance between highlighting scientific discoveries, 'busting' popular myths, and providing realistic tips that can be applied to everyday life. The stories he shares about his experiences and the brain's ability to adapt and heal are incredibly inspiring."

— Sandy Nunez, ESPN

"Genius is the ability to explain the complex in simple terms. In this wonderfully crafted book, Professor Rahul Jandial has explained the structure and function of one of the most complex, least understood organs in medicine and science — the brain . . . He provides detailed and moving accounts of real-life case studies from his own practice to explain the intricate workings of the brain while, at the same time, dispelling myths and providing practical ways in which we can all improve the function of our own brain. A must-read book to enhance brain performance now and long into the future. Dr. Jandial is a genius."

— Olympian Greg Whyte

LIFE LESSONS FROM A BRAIN SURGEON

Practical Strategies for
Peak Health and Performance

RAHUL JANDIAL, MD, PhD

HARVEST
WILLIAM MORROW
Boston   New York

First Mariner Books edition 2020
First published in 2019 as *Neurofitness: A Brain Surgeon's Secrets to Boost Performance and Unleash Creativity*
Copyright © 2019 by Rahul Jandial
Illustrations copyright © 2019 by Annie Hurley

www.harpercollins.com

Library of Congress Cataloging-in-Publication Data is available.
ISBN 978-1-328-96924-8 (hardback) ISBN 978-1-328-96983-5 (e-book)
ISBN 978-0-358-41095-9 (paperback)

Book design by Allison Chi

Printed in the United States of America
24 25 26 27 28 LBC 8 7 6 5 4

To my sons

CONTENTS

PROLOGUE

I t felt absolutely medieval. This wasn't a maneuver that would work with gradual pressure, like the tightening of a vise. It needed a quick, crushing force. So I used a head holder, with one-inch-long steel pins, to secure the skull to the operating table. That way, if my patient started to move, her head would remain still, and I wouldn't accidentally kill her.

The three metal pins would need to bite down into her skull after puncturing her scalp: one pin in her forehead, two in the back, all connected to a C-shaped clamp. While my assistant held up the patient's head from the neck, I explosively captured her cranium inside the steel device. The jarring noise from the metal gears made the students, nurses, and doctors standing behind me in the operating room fall silent. The first of several hundred steps that needed to go smoothly, quickly, and perfectly had just been completed.

So began my first time opening the skull of a living human being. I was a third-year resident at the University of California, San Diego, Department of Neurosurgery. My patient was in her midthirties and had come to the hospital's emergency room two days before, reporting a peculiar weakness and awkwardness in her left arm and hand. An MRI had revealed a bright white abnormality on her brain — a tumor the size of a peach.

Many times before, I had stood beside senior neurosurgeons, assisting, observing, and learning. But this was my first time going solo.

It's an odd thing—brain surgery. There's fear, of course, but also awe that you're literally inside somebody's head, which elicits intensity as well as excitement. I don't want to sound indelicate, but for me it's a thrill. Some people like skiing, or mountain climbing, or playing poker. I like operating on people's brains.

The risk is that I will nick a vein and a part of the brain will die. Or I will go in at the wrong spot and won't be able to reach most of the tumor. Or everything will seem to go perfectly during the surgery, but the patient will wake up unable to speak for the rest of their life.

The hope—and why I do it—is that this woman, who just got married three months ago and has much of her life ahead of her, will have her strength and fine control of her left hand restored as good as new.

Despite having abnormal tissue in her brain, this patient was pretty lucky because the mass wasn't malignant. Her life was not at risk from the tumor, just from me. But as long as the tumor remained and continued to grow, her muscle weakness could worsen and spread. It was nestled in the motor strip of the right parietal lobe—a half-inch-wide, seven-inch-long ribbon of brain tissue that sends movement signal to the left side of the body. This particular type of tumor is called a meningioma because it grows from the lining (*meninges*) of the brain. Since the skull can't stretch, the tumor knuckles into the brain, deforming it, without actually penetrating the tissue. The pressure, however, interferes with the electrical signals, leading to weakness.

After drilling off a circular piece of bone near the top of her skull —what brain surgeons call "turning the flap"—I gently sliced with a number-11 scalpel into the dura—the thin, cloth-like membrane that protects the brain. I scored and lifted the dura but went no farther.

And there it was. I could see the tumor on the very surface of the brain. In contrast to the glistening opalescence of healthy brain tissue, it was yellow, dull, and irregularly spherical.

I began by entering the center of the tumor, coring it like the yolk of a hard-boiled egg until it was hollow, leaving behind only its tougher rim. Then, I delicately worked its shell away from the surrounding brain, collapsing it into itself. This is the hard part, because the edges have bridging spider-silk-thin fibers, and the surrounding tissue is as soft as pudding. Slowly, methodically, I divided those wisps with a curved eight-inch scissor.

Two hours of doing this under magnification and illumination, and the tumor was out. I bathed the brain's surface with sterile water to check for any oozing or dripping blood vessels. Then it was time to close through reverse maneuvers. I reattached the bone flap to the rest of the skull with a thin titanium mesh and tiny plates and screws, stitched the scalp back in place, and finally removed the clamp holding her head still.

Three days later, when her brain was no longer stunned by my invasion, she had full strength back in her left hand and arm, and I knew what I wanted to be great at.

Fifteen years and thousands of operations later, brain surgery is still a thrill unlike any other. My three sons tease me about having gone to school until the thirty-second grade — literally twenty years beyond high school — but that's what it took to become a brain surgeon and also add a PhD in neurobiology. Even so, I feel like I've only just glimpsed at the mystery and potential of the human brain. It is my obsession.

These days I not only perform brain surgeries, I also teach medical and graduate students to conduct neuroscience and oncology research in my laboratory at the City of Hope, a cancer treatment and research center in southern California. I travel to countries such as Peru and the Ukraine on surgical missions. I have written ten academic books and more than 100 papers about brain surgery and neuroscience used by medical students, PhD students, and neurosurgeons.

But there's something eating away at me that no amount of surgery or science is going to fix. It's a kind of infection of the mind spread by close contact with television, websites, sensationalist books, and certain companies eager to sell the public on simplistic pseudoscience and nonsense.

Perhaps you've heard claims like this:

- SOME PEOPLE ARE MORE LEFT- OR RIGHT-BRAINED. I explain how this myth was concocted.
- THE GUT IS A SECOND BRAIN. Not really. The brain does project nerves out of the skull to nearly every millimeter of your body, including an extensive network of nerves in your guts that monitor your gastrointestinal tract. But many patients have undergone multiple variations of near-total bowel removal without demonstrating any attributable mental dysfunction.
- BRAIN TRAINING IS BOGUS. In truth, leading researchers at top universities around the world continue to investigate the effects of computerized "brain games" and all sorts of other training methods to improve cognitive performance.
- MEDITATION IS NOT BACKED BY HARD SCIENCE. False. A recent groundbreaking study directly measured the mind-calming effects of meditative breathing, elegantly showing the practical physiology underlying this ancient ritual and now modern practice.

These days it's harder than ever to sort the facts from the phony claims.

And many of these ideas, peddled by self-declared experts, could be holding you back from reaching your real personal potential. I've treated patients who truly believed that herbs or meditation would cure them of their brain cancer, and thus they delayed life-saving surgery. I've met people whose strokes might have been prevented had they followed a few simple rules of staying neurologically fit. I've known students in my medical school classes who thought they would get better grades by taking "smart" pills, which, in truth, only allowed them to work longer and harder at being just as stellar or mediocre as they were to begin with.

This book is my attempt to separate the BS from the brain science, the hype from the hope. I want to help you achieve your goals and ensure that you and your loved ones never end up on my operating table.

To that end, I make no claims that are not backed by current hard science. I neither minimize the risks of alternative medicine nor exaggerate the benefits of traditional Western-style medicine. Knowledge is a moving target, and I share what we know now and what we hope to find out.

The wonders of the brain require no exaggeration. Between our ears live an estimated 85 billion neurons—as many brain cells as there are stars in the Milky Way galaxy. Each of those neurons has thousands of thread-like connections, called synapses, linked to other neurons in the brain—more than a hundred trillion connections. That's ten times more than the estimated number of galaxies in the entire universe. The brain's complexity is unparalleled and vast.

Even when brain surgeons know that a particular procedure works to relieve suffering, we often don't know why. I can implant an electrode deep into your brain that I know will relieve depression or OCD or improve Parkinson's disease. How? Brilliant question. When you find out, let's connect.

One thing we brain surgeons do know for certain is that every brain can make a comeback following a devastating illness or injury. We witness the living proof in our patients who have experienced strokes, trauma, or brain cancer and who manage to make incredible recoveries. They relearn to walk and talk, regain fine motor skills, and improve their cognitive functioning using techniques that can and must be practiced not only in a hospital but at home. If my patients can do it, why would anyone doubt that healthy people can push their cognitive power into a higher gear?

To help you get there, I have packed this book with practical, tested, real-world strategies and hacks to achieve peak performance from a brain-centric approach to diet, creativity, sleep, memory, and so much more — for young or old, healthy or ailing.

Don't worry, I am not going to tell you to put down your smartphone. Devices aren't going anywhere, and they are not inherently bad. It all depends on how they are used. In fact, my patients often use devices during their brain rehab, and I'll show you how to use your digital tools to keep your brain sharp and agile.

In this book, I take you on a journey into the operating room, around the world on my surgical missions, and inside my research lab so you can see what it's like to be on the frontlines of brain science. I venture to the outer edges of frontier neuroscience to reveal the latest, most important brain breakthroughs that are turning science fiction into reality, and I share the stories of some of my patients who have made remarkable recoveries.

Each chapter includes one or more of these special sections:

- NEURO BUSTED, where I address popular myths and misunderstandings;
- NEURO GEEK, where I dive a little deeper into cool (if wonky) scientific theories, discoveries, and history; and
- NEURO GYM, where I boil down the science to actions you can apply to your own life.

You'll find solid, state-of-the-art information, and you won't have to follow an exhausting and time-consuming regimen to see results. As a surgeon who works with patients on a daily basis, and as a dad with three boys and a wife who is a cancer scientist with her own rigorous schedule, I know that life can get in the way of your best intentions.

If I give a patient a list of ten postsurgery recommendations, I know that 95 percent of them won't follow through on all ten, so I point to the two or three items on that list that are the highest yield. I will do the same for you here, focusing on the brain-building strategies that won't waste your time.

I've been waiting to write this book for a decade, until a stage in my life when I am no longer a rookie but far from a retiree.

I hope you find it worthwhile.

Rahul

1

AN ANATOMY LESSON
LIKE NO OTHER

I hated Anatomy 101. The basic class for all incoming medical students, mine took place in a giant room reeking of formaldehyde and crammed with naked corpses on steel tables, each one surrounded by a coven of students eager to begin digging into it.

I found the whole thing gruesome and repellent yet somehow simultaneously dull. Where was the risk in dissecting a cadaver? The whole thing was unsettling, so much so that I never once held a scalpel during that entire first year, insisting instead on merely observing as other students cut and explored. Surgery, apparently, was not in my future.

Even the brain, upon first meeting, proved to be as much a disappointment as the rest of the body. For all the lectures and textbooks extolling its wonders, the thing I saw that first year of med school — dead and bloodless — looked like a beige, corrugated cauliflower. I could see why the ancients ignored it for millennia. The one thing that caught my attention about it was how hard the thing was to get to. To penetrate the skull, we were issued an ordinary electric saw from a hardware store and told to make a circular cut around its circumference.

My disinterest, even disdain, for human anatomy disappeared in my third year of medical school, the first time we were permitted to observe heart surgery on a living patient. The intensity, the stakes, the adrenaline were what I had been waiting for. Until then, I had been seriously questioning whether medicine was for me. It had all been books and boredom and dead bodies. Now blood was flowing. I knew I wouldn't be able to spend my career only writing prescriptions. Horrible as it sounds, I needed to get my hands bloody.

After completing four years of medical school at the University of Southern California, I was accepted into a general surgical internship at UC San Diego, intending to become a cardiac surgeon. The heart seemed like the gnarliest of surgical specialties. Neurosurgery never entered my mind; I hadn't observed a single brain surgery in four years of medical school.

That first year of internship, we would-be surgeons rotated, month by month, through each specialty, from trauma and orthopedics to plastic surgery, abdominal, heart, ear-nose-throat, and, supposedly, the brain. But we were considered such rubes that the neurosurgeons never even allowed us into the operating room, keeping us outside to serve as glorified scribes in the pre- and postoperative areas.

At the end of that year, however, a rumor started floating through the hospital's hallways that the neurosurgeons were about to fire their own chosen intern; for some reason, he wasn't working out. That subspecialty was so elite that they took only a single trainee per year, compared to the three or more in every other specialty.

One evening, a neurosurgery resident sat down next to me in the hospital cafeteria and mentioned that his department couldn't run without that one resident per year.

"They're looking to grab someone from the other surgical services," the guy told me.

"Who are they thinking of hitting up?" I asked.

"They're thinking about you," he said.

I thought: *Really?*

I knew the least about the brain. That is one of the areas that surgical interns ignore if they're not planning to specialize in it. You just don't waste your time on it because, down the line, if you ever have a case that involves the brain, you punt to a specialist, no questions asked.

"You have a reputation," the resident told me. "You know the least but get the most done. They like how you work and how you don't flinch. The professors' main concern is whether you'll have enough time to catch up on the knowledge base and pass the exams. They know you have the hands — the cardiac surgeons told them — but they wonder about your smarts."

"Thanks, I guess," I said, unsure how to respond.

Within a week, the professors and I sat in a meeting to discuss their formal offer to switch to neurosurgery.

"Why don't you give it a try," one of them said. "If you don't master the content, we'll fire you."

He laughed. The others laughed, too. They weren't kidding.

"I've never even seen a brain surgery," I told them. "Before jumping ship, I'd like to see one."

They offered for me to observe a bifrontal craniotomy scheduled for the next morning. The operation, they said, begins with removing most of the skull over the forehead.

"And you can do that without killing the patient?" I asked.

They laughed again at my naiveté.

But there was no more laughter the next morning, at 7:30 a.m., when I stood across from the surgeon on the other side of an operating table. His patient, lying before us, was covered by a sheet except for the top of his head, which had already been shaved. The surgeon cut the scalp, drilled and cracked open the bone, incised open the dura, and revealed undulating white flesh speckled with tiny blood vessels. For a moment, it felt like a violation. Heart surgery is impressive, but in some ways it's like working on a car engine: it's all valves and pistons and fuel lines. But the brain is different.

Here at the mysterious core of human being, I thought maybe a living person's skull should be sacred space, taboo to enter.

That feeling lasted about five seconds. Then came the thrill. If the cranial vault is a sacred sanctuary, so be it: I could become one of those few trusted to enter it. Later that day I let the professors know: I would accept their offer to enter neurosurgical training.

So began my anatomy lesson like no other. Now please allow me to show you inside my workshop.

BEYOND THE SKULL

To begin: the brain doesn't actually sit inside the skull; it floats, protected by a natural shock absorber called cerebrospinal fluid, or CSF for short. CSF is produced at a rate of about two cups per day from inside the brain's deep, hidden chambers: the ventricles.

Although it looks like water, CSF is filled with bioactive ingredients that serve as the brain's "nourishing liquor." It carries bioactive factors that keep the brain fit and also drain away the brain's waste products.

A weird thing about the brain is its peculiar texture when you touch it. You might assume it's like a muscle, or body fat — that you could push your finger into it as you would into someone's belly, and it would smoosh in a little and then rebound. But that's not how it works. In reality, the texture of the brain is like no other flesh. It's more like flan or bread pudding. Push your finger against it, and your finger will sink straight in. Take a thimble and you can scoop up an easy million or so brain cells.

Those cells on the outer lining of the brain, by the way, are the most precious ones. You have probably heard of the cerebral cortex. That's not just a synonym for the whole brain. The word *cortex* comes from the same Latin word that gives us *cork*, which grows on the bark of a type of oak tree. The cerebral cortex is the bark on the brain — incredibly, less than one-fifth of an inch thick — where most of the magic happens in human beings: consciousness, language, perception, thought.

Visually, the most notable feature of the brain's surface is how it looks like tightly tucked hills and valleys. Each of the little hills is called a gyrus, with a soft "g." (So it's *"jie*-russ," not "guy-russ.")

The valley or sunken section is called a sulcus, which is pronounced with a hard "c." (So it's "sulk us.")

The reason for all this folding is that it permits a greater surface area. Unfolded, the cerebral cortex would be the size of an extra-large pizza. The brain wants the most acreage it can get of that thin but powerful cortex, so it figured out a way to fit more of it into the skull, by folding it like an accordion or a pleated curtain.

What you have to understand about the cortex is that it's all "gray matter," the central bodies of brain cells. Under a powerful microscope, these neurons appear to line up vertically like pine trees in a forest. And like the roots of a tree, each neuron has a branched network of thread-like connectors that link to other neurons. These connectors — the biological equivalent of cables — are the "white matter." Sixty percent of the brain is white matter.

The incoming fibers that carry messages *from* other cells are called dendrites. The outgoing fibers that send messages *to* other cells are called axons. So if one neuron wants to talk with another neuron, it sends an electrical signal down its axon to meet one of the other neuron's dendrites. But they never physically touch. Think of Michelangelo's painting in the Sistine Chapel of God and Adam's fingers reaching toward each other.

The space between, called a synapse, is where a variety of chemical messages swirl. Those chemical embers, called neurotransmitters, float across the synaptic cleft. There are dozens of these neurotransmitters — some that you might have heard of include dopamine, serotonin, epinephrine, and histamine — and they all have different effects on neuronal communication and function. Put it all together, and you can begin to understand the design that can generate the infinite variety of feeling, thought, and imagination that humans experience.

NEURO BUSTED: BRAIN CHEMICALS PLAY MANY ROLES

Some people think of dopamine as the "feel good" neurotransmitter, the one that showers your brain when you are overcome with feelings of love or happiness, or activates with the help of drugs like cocaine. But like all neurotransmitters, it has multiple functions. Yes, dopamine is closely involved in generating subjective feelings of pleasure. But a lack of dopamine in the brain causes Parkinson's patients to struggle with movement. And when medicines like L-dopa are given to replace deficient dopamine in order to address movement issues, the range of possible side effects are quite revealing. Some patients develop a gambling addiction, and some can become hypersexual. The bottom line is that assigning one feeling or cognitive function to each neurotransmitter is a gross oversimplification. All neurotransmitters — including not only dopamine but also epinephrine, norepinephrine, glutamate, histamine, and many others — play different roles in different parts of the brain.

But let's get back to the large map of the brain. Functionally, the cortex is divided into four sections, or lobes, each devoted to a particular set of tasks. Structurally, though — seen from the top — the brain is also divided into a left half and a right half. Connecting the two halves, deep inside the brain, well below the cortex, is the corpus callosum (Latin for "tough body"), a bundle of hundreds of millions of axons. Each of the four lobes, and all the other brain structures located deeper in the brain, exist in pairs, just like your eyes, your ears, and your limbs.

Let's start with the lobe most unique to humans: the huge frontal lobe that bulges beneath our foreheads.

FRONTAL LOBE

The frontal lobe plays a primary role in motivation and reward-seeking behaviors.

When you're carefully paying attention to what your teacher or boss is saying, that's your frontal lobe at work. Doing math? Frontal lobe. Crossword puzzle? Frontal lobe. Trying to figure out how to handle a former friend who has lately been talking behind your back? The integration of all those feelings, memories, and possible responses requires the quarterbacking of the frontal lobe.

And when you feel like jumping out of your car and screaming at somebody in a traffic jam, it's the frontal lobe that's supposed to step in and say, "Hold up, not worth it."

Actually, that kind of complex decision-making and juggling of conflicting possibilities is managed by a section of the frontal lobe called the prefrontal cortex, or PFC. Just like it sounds, the PFC is the most forward-facing part of the frontal lobe. This is where some of our most human faculties lie: planning, personality, rule learning, and other "executive" functions that permit us to live in a complex, nuanced world pummeling us with stimuli.

Another subsection of the frontal lobe, near the outside edge of your eyebrows, can be found only on one side of your head: the "dominant" hemisphere, usually meaning the left side (if you're right-handed) but very rarely on the right side (even if you're left-handed).

This section, called Broca's area, is where your ability to speak resides. In chapter 3, "The Seat of Language," there is a full description of Broca's and other nearby areas that control not only the ability to speak but to understand.

PARIETAL LOBE

Running just a few inches from the crown of your head back toward the nape of your neck, the parietal lobe controls sensation. In the first half of the twentieth century, a Canadian-American neurosurgeon,

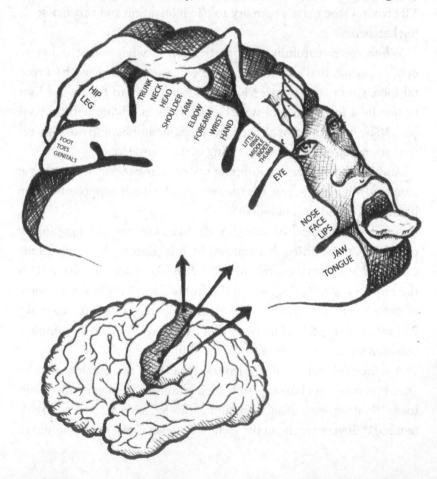

Wilder Penfield, mapped exactly which parts of the parietal lobe correspond to which parts of the body. Using a tiny probe with a forked tip and a minuscule current running between its tines, he tickled the parietal lobes of living, conscious patients who were undergoing brain surgery.

(It may sound arcane, but we still use awake brain surgery to help patients. It turns out the surface of the brain doesn't feel. The scalp feels pain, but the surface of the brain has no pain receptors. It relies on its emissaries, the nerves it sends out into the face and the body via the spinal cord. So, if I numb your scalp and open your skull while you're under, and then taper off the anesthesia, you will wake up groggy, pain-free, and able to let me know if I'm touching something that interferes with your ability to move, talk, see — or anything else.)

Step by step, Penfield systematically marched up and around the parietal lobe to identify the corresponding feeling of touch, throughout the body. This area felt like someone touched my foot; that area felt like the stroking of a cheek. Penfield surveyed the cartography of the parietal lobe to generate what is today known as the cortical homunculus, or "little human."

Notice that just your tongue, lips, and fingers get about as much area in the brain devoted to them as the entire portion of your body below your thighs. No wonder a kiss or caress stirs us.

Amazingly, more than forty years since Penfield's death, his maps remain so accurate that we still use them today as a general guide to where the motor and sensation functions are located.

OCCIPITAL LOBE

The portion of your brain at the very back of your head is called the occipital lobe, from the Latin roots *ob* ("behind") and *caput* ("head"). It's the brain's visual processing center. An injury or stroke to *both* the left and right occipital lobes causes blindness, even though the eyes work fine.

Where things get weird is if you injure only the left occipital, or just the right. Depending on where the damage occurs, sometimes the effect on vision is barely noticeable. Occasionally, however, the person develops a manageable condition called homonymous hemianopsia: partial blindness in *both* eyes but on just the left side of their visual field, or just the right side. They can see fine looking straight ahead, but their peripheral vision is shot.

TEMPORAL LOBE

Place a finger one inch above each of your ears. Just below that spot are the two halves of the fourth lobe: the temporal (as in the *temples* of your head). Not surprisingly, they handle the processing of sounds in general and the understanding of speech in particular.

Dr. Penfield also used his electric probe to tickle the temporal lobe. When some spots were stimulated, he found, a person would suddenly be unable to understand speech. In other spots, they would have an astonishing variety of sensations: of dreamlike states, of suffocation, burning, falling, déjà vu, even profound spirituality.

I once used an electrical stimulator on the temporal lobe of a patient who had a tumor deep within it. Seeking to find a spot where I could safely dissect deeper, I stimulated here and there, each time asking what, if anything, he was experiencing.

"Listening to Kendrick Lamar!" he said at one point. "Kendrick's rapping!"

It was so vivid, the man told me, it sounded like I'd turned on a speaker next to his ear.

DOWN UNDER

The four lobes we discussed are just the lobes of the cortex, the outermost layer of the brain. Underneath, the axons and dendrites connect and channel the billions of neurons above to each other and to

NEURO BUSTED: GRAY MATTER ISN'T GRAY

In the living brain, gray matter isn't gray, and white matter isn't white. Those colors appear only in dead brain tissue filled with preservatives. Inside a living brain, "gray" matter is actually a shimmery beige-pink; "white" matter — the axons wrapped with a fatty myelin sheath — is the color of a glistening pearl. Under the bright lights of the operating room, the iridescent surface of the brain is densely punctuated with ruby-red arteries and hyacinth-blue veins.

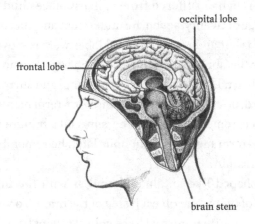

occipital lobe

frontal lobe

brain stem

deeper brain structures below. These subcortical structures — those beneath the cortex — serve in part as transit hubs for signals coming from and going to the spinal cord. They modulate and fine-tune those messages.

The hippocampus sits in the basement of the temporal lobe. Its name derives from its resemblance, first noted by a sixteenth-century

Venetian anatomist, to a sea horse (from the Greek *hippos*, meaning "horse," and *kampos*, meaning "sea monster"). There are actually two, below both the left and right temporal lobes, and they are essential for forming new memories.

Usually, either your left or right hippocampus is dominant, so this redundancy allows us to remove one temporal lobe if you're having epileptic seizures emanating from it without destroying your ability to remember new people, places, and events. We figure out which side is most dominant by shutting down one side at a time with a temporary paralytic agent, and then we ask you a series of memory questions.

Scientists first learned the role of the hippocampi due to the tragic case of a man known to the public only as H.M. until after his death in 2008. Henry Molaison had suffered from epilepsy since childhood until, in 1953, at the age of twenty-seven, he underwent an experimental surgery: Parts of his left and right temporal lobes were removed in hopes of eradicating the abnormal electrical discharges causing his epilepsy. They ended up taking *both* of his hippocampi and surrounding regions. Afterward, he could still form brief, short-term memories (he could, for instance, remember what a person said a minute ago), but not any new long-term memories (an hour later, he remembered nothing of the conversation).

The amygdala is shaped like an almond, and as with the hippocampi, there are two of them, one on each side of the brain. To visualize where they are located, imagine two lines going straight back from the eyes, each one intersecting a third line going between the ears.

Unfortunately, this paired structure has gained infamy as the place where fear lives. This tremendous oversimplification is naive and misguided. It arose from press descriptions of a rare disease called Kluver-Busy syndrome, which is marked by a near-total loss of fear after injury to the amygdalae. And while it's true that the amygdalae do play an important role in fear, they also do so for other positive emotions. It's not the fear center, but it is an intense emotion hub.

The thalamus is another paired structure. It's larger than the other deep-brain structures and sits near the bottom of the brain, atop the brain stem. In the true center of our semispherical brain, it is a large cluster of gray matter that serves as a kind of train station for all the axons passing through on their way to the spinal cord. Here the axons have their signals tweaked; signals being sent to move the muscles are smoothed and refined. Sensations coming *from* the body, meanwhile, are likewise modulated and sent on the right tracks toward the appropriate corners of our cortex. The thalamus is like the old-school switchboard operators who direct a stream of incoming and outgoing calls to the right place.

The hypothalamus, located directly below the thalamus, is only the size of a plump grape, but it regulates hormones that control blood pressure, body temperature, growth, and more. It is a no-fly zone during surgery.

The brain stem lies below everything else, a structure at the bottom and center of the brain that is no thicker than your thumb. If you stuck your finger in your mouth, it would be pointing at the brain stem. From behind, it's about even with a shirt collar. The brain stem is the portion of the brain that controls basic functions like breathing, sleeping, heart rate, consciousness, and pain sensitivity. Damage it and recovery is impossible. Miracles don't happen after this area is injured.

The cerebellum (Latin for "little brain") lies behind the brain stem, below the rest of the brain, and can be found in all vertebrates. It helps to refine your physical movements, with a particularly strong effect on coordination and timing. Although it was once believed that motor control was the cerebellum's only function, neuroscientists now understand that it also plays an important role in a variety of mental and emotional functions. Some now think of it as a "supervised learning machine" that refines thoughts and emotions just as it does movements. As with much of our knowledge about the brain, it's still TBD.

BELOW THE NECK

The brain is always depicted as an isolated organ sitting atop the body like some master controller. In truth, its tentacles reach throughout the body. Thirty-two nerves branch out of the spinal cord, the tail of the brain, and they exit the spine and go into your arms and legs, allowing your brain to sense what your fingertips touch, telling them in turn whether to pick up that grape or toss that stem. Other nerves arise directly from the brain and descend to your heart and your guts to modulate their function, telling them how quickly to pulse and writhe — and, in turn, telling you when you're feeling so nervous you have "butterflies."

The brain's reach into your body is not only mediated by nerves. The deeper structures in the brain, like the hypothalamus, make master hormone regulators that trigger the nearby pituitary gland to drip hormones into your blood. As these hormones descend from the blood in your brain to your body, they tell the thyroid, adrenals, testicles, and ovaries what to do. All the glands in your body are under the dominion of chemicals released by the pituitary, which hangs underneath your brain just behind the upper bridge of your nose. As with nerves that the brain sends down, hormone levels are detected in reverse order to keep us finely tuned. Disruption is when diseases ensue.

We still have no clue how consciousness arises from flesh and blood or how the mind arises from matter. We early cartographers of the cranium have made only the roughest of maps. And I can't wait to find out how the gaps will be filled in.

NEURO GEEK:
THE SECRETS OF EINSTEIN'S GLIA

Albert Einstein left clear instructions on what to do with his remains in the event of his death: he wanted to be cremated, his ashes to be scattered secretly. Instead, when he died on April 18, 1955, the pathologist on call, Thomas Harvey, stole Einstein's brain and took it home. He sliced it into 240 pieces, put them in a preservative, and then placed the pieces in two jars in his basement. Eventually, he sent some of the slices to scientists around the world.

One of those scientists was a neuroanatomist named Marian Diamond. I remember Dr. Diamond well from her popular anatomy class at Berkeley. She was the first scientist to show, in rats, that an enriched environment, in the form of toys and other rats, increased the thickness and performance of their brains. But her greater fame came in 1985, when she reported the results of her study of four slices of Einstein's brain.

What she found was that he had significantly more glia — the often ignored brain cells that surround and protect neurons — than in the average male brain. This opened the door to our understanding of the importance of glia as more than mere bystanders in the brain's development.

We now understand that the approximately 85 billion glia feed neurons with nutrients and oxygen, insulate them from each other, destroy invading pathogens, remove dead neurons, and enhance their communication.

2

BEYOND MEMORY AND IQ

After the second year of medical school, all would-be physicians must take a grueling day-long test: Step 1 of the United States Medical Licensing Exam. It's an eight-hour multiple-choice test on anatomy, biochemistry, behavioral sciences, genetics, immunology, pathology, pharmacology, physiology, plate tectonics, quantum mechanics, rocket science, and the Upper Paleolithic history of Siberia.

Some of that was exaggeration, but it is painful, because you're scored against about 18,000 other medical students who all excel at memorizing and taking tests.

It's not quite true that a physician's entire future depends on their score, but it's close. That score is the chief element determining not only how prestigious an institution they can train at for residency, but also the speciality for which they can train.

Preparation requires literally hundreds of hours of memorization. I scored better than most, but there was a guy at my residency rumored to have gotten the highest score in the country that year. He was the same guy I mentioned in the first chapter, the one initially selected to become the neurosurgery resident at our hospital who was then kicked out after just a few months.

He did not lack for IQ. He was great at acing tests. But, from what I've been told, he didn't know when to ask for help with a dangerously sick patient or when he could handle one on his own. The constant crises left him flustered. Juggling a neurosurgery service with twenty patients, as a rookie surgeon, requires multitasking and judgment—very little to do with multiple-choice questions and raw knowledge.

The best scorers on tests in medical school are given the opportunity to train in surgery, but they are never evaluated for technical ability and performance under pressure. So, as you can expect, there is often a mismatch of smarts and ability.

Of course, intelligence matters. The question, however, is *how much* it matters. Bill Gates and Oprah Winfrey didn't become giants in their fields without intellectual firepower. But then again, they didn't succeed at turning their smart ideas and insights into business juggernauts by IQ alone. They needed good judgment, determination to succeed, and the ability to manage, delegate, and inspire those around them.

Let's take a look at how *all* those essential capacities emerge from the brain and how to maximize your own natural gifts.

NEURO GEEK: THE FLYNN EFFECT

In 1984, a New Zealand academic named James R. Flynn made a curious discovery. While examining IQ scores going back to the early twentieth century, he saw that the average had been steadily rising by about three points per decade. The middle-of-the-pack score back in 1920 would now be graded as 70 — which by today's standards would be considered mildly intellectually impaired. The average person taking the test today would have been given a score of 130 — nearly "genius" level — had they been graded back in 1920.

Some researchers explain away the so-called "Flynn effect" by arguing that people have simply gotten better at taking tests. Improved teaching methods, the introduction of kindergarten and preschool, and higher graduation rates have all resulted, they claim, in young people simply testing better.

Flynn, however, insists that kids are literally getting smarter in part because of school, but also because of better nutrition and fewer childhood illnesses than their grandparents experienced. Even more important, Flynn says, is that our world has become ever more cognitively challenging. A hundred years ago, nearly one-third of Americans lived on a farm; today that figure is less than 2 percent. Only in the 1920s did radio become popular; not until the late 1950s did television reach most homes; as recently as the year 2000, fewer than half of all U.S. homes had access to the internet; the smartphone as we know it didn't exist until Apple released the iPhone in 2007.

"We are the first of our species to live in a world dominated by categories, hypotheticals, nonverbal symbols, and visual images that paint alternative realities," Flynn has written. "We have evolved to deal with a world that would have been alien to previous generations."

The undisputed fact of the Flynn effect demonstrates that human beings are capable of getting smarter, that intelligence is not determined simply by DNA.

HOW YOUR BRAIN REMEMBERS

For much of the twentieth century, scientists believed that each memory in the brain was stored as a web of connections between neurons — not in any single neuron or collection of neurons, but in how they linked together. This view was seemingly proved beyond doubt in a famous paper published in 1950 by psychologist Karl Lashley. He performed hundreds of experiments on rats in which he taught them to remember a maze, a task, or an object and then made surgical incisions in various spots of their brains. No matter where he cut, the rats still remembered what they had learned — but just a little less well. Two incisions made them a little more forgetful; three made them even worse, and so on. No one spot mattered more than any other.

"There are no special cells reserved for special memories," Lashley wrote.

That orthodox view first came under attack in 1984, when neuroscientist Richard Thompson trained rabbits to blink whenever they heard a particular musical tone by repeatedly pairing the tone with a puff of air to their eye. Once the rabbits had learned to associate the musical tone with the puff of air, they blinked every time they heard the tone, even without the puff. But then, contradicting Lashley, Thompson found that if he removed just a few hundred neurons from a section of the cerebellum near the brain stem, the rabbits no longer blinked. Somewhere in those neurons, he described, the rabbits' memory linking the puff of air and the musical tone had been stored.

By 2005, scientists were showing that individual neurons are involved in recognizing particular faces. Upon being shown a picture of Jennifer Aniston, for instance, a single neuron in the hippocampus reacted. Another lit up in response to pictures of the actress Halle Berry.

Since then, researchers have developed an array of neuromolecular tools to create false memories in mice — the sort of thing seen in the Christopher Nolan film *Inception*.

Other techniques have been used to make the fear associated with a particular stimulus disappear — a treatment that could one day be of value to people suffering from phobias or post-traumatic stress disorder.

LEARN FROM THE GERM

Memory of one kind or another is at the heart of all life. What else is DNA but a way for life to remember its own blueprint for reproduction?

You might assume that a brain is necessary for remembering, but that's not so. Consider *E. coli*, the single-celled bacteria that live in our guts and the guts of most other warm-blooded creatures, where they are usually harmless but occasionally cause food-borne sickness. Incredibly, they have a version of short-term memory. When they're swimming around in your intestines, looking for food, they will keep going in a more or less straight line as long as they find nothing worth snacking on. But once they find something nutritious, they will stop, eat, and then pirouette from that spot, turning in a tight circle in hopes of finding more deliciousness close by. Once they run out of treats in that localized area, they will then continue on their way.

Almost all animals do this: It's called area-restricted search. If a pigeon finds a single crust of bread under a chair, it will keep pecking for other crumbs nearby until nothing is left. Then it flutters off in search of another spot.

Both parts of this strategy are really important: making sure you find every last crumb in a given area, and then systematically searching other areas.

What's really strange is that memory works the same way: by area-restricted search. If I ask you to list all the animals you can think of, you are likely to begin with the category of "pets" and list cats, dogs, goldfish, parakeets. Once you run out of items in that category, you'll move on (like the pigeon who can find no more crumbs) to another category: farm animals such as cows, chickens, pigs, goats, and horses. When you can't think of any more in that category, you switch to jungle animals: lions, tigers, monkeys. And so on. The same process by which *E. coli* looks for food in your gut is how you try to remember that last grocery item you were supposed to buy. (Is it in dairy? Fruits and vegetables? Meats?)

A really cool study of this was published in the journal *Memory and Cognition*. It found that smarter people can list more animals overall than less intelligent people, but only because they are better at thinking up *more categories* in which to mentally search. When the researchers ran the test again with another set of participants, they required the participants to use a set list of categories provided (pets, farms, jungles, forests, etc.). As a result, the gap between the smarter and less smart people disappeared.

On the other hand, people with early signs of dementia tend to show the opposite problem: When trying to recall a long list of items, they are less good at going through each category. They move to another category before exhausting the first.

So when trying to remember stuff, try to follow the lesson of *E. coli* and pigeons: Intentionally practice area-restricted search. Diligently scour your brain first for categories and then for items in each category.

For instance, let's try an easy exercise that will take you less than five minutes. First, take a paper and pencil or open a new document on your computer, and set a timer for *two minutes*. When you are ready, write down the names of as many kinds of water-dwelling creatures as you can think of in the time given.

Ready? Begin!

Okay, once you are finished, I would like you to try it again. This time, however, I want you to use the following categories, again giving yourself just two minutes to list as many as you can.

When you're ready, begin!

1. Freshwater fish
2. Ocean-dwelling fish
3. Ocean-dwelling mammals
4. Dangerous fish
5. Sea creatures that have shells

If you gave it a shot, the list of five categories helped you to think of new kinds of water-dwelling creatures. Area-restricted search works for people just like it works for germs!

THE LEARNING TREE

An amazing series of experiments by Monica Gagliano, an evolutionary ecologist at the University of Western Australia, has shown that a plant can learn.

The first plant Gagliano studied, *Mimosa pudica*, is a perennial herb in the pea family. Popularly known as the "sensitive plant" or "touch-me-not," it is famous for having leaves that fold inward and droop whenever it is touched or shaken. After a few minutes, the leaves reopen.

Gagliano decided to test whether the plant could learn to ignore a particular type of disturbance. She placed dozens of them in holders that would periodically allow them to drop down by about a foot. At first, the plants' leaves immediately folded inward after the fall. But after repeated descents, the leaves stopped reacting; they remained open. Apparently, they had become habituated to the drops. They had learned.

Then, in 2016, Gagliano published an even more astonishing report. Most plants grow toward the light, right? She designed an experiment to see if plants could learn a conditioned response in the same way that Pavlov had taught dogs to salivate in response to the ringing of a bell. This time she used forty-five seedlings of another kind of pea plant, *Pisum sativum*, and placed fans and a light source either on the same side of the plants or on the opposite. After three days of this, she tested the plants' growth on a fourth day, when only the fan — without any light — was turned on.

Sure enough — without a brain! — a majority of the pea plants began growing toward or away from the fan, exactly as they had been taught to expect light to appear based on their three days of training.

Gagliano and others have put forward some clever hypotheses for how plants can learn such tricks, but for now, this much is certain: the ability to learn and remember is so important to life, even plants and bacteria do it!

NEURO BUSTED: IS BRAIN TRAINING BS?

Plenty of news articles these days claim that brain training is junk science, so when the most popular commercial provider of on-line brain games, Lumosity, was fined $2 million by the Federal Trade Commission for making unsubstantiated claims, the media pounced.

But as a neurosurgeon and neuroscientist who has studied the scientific literature and seen the benefits of brain training on my own patients, I know that at least some kinds of brain training — not Lumosity, perhaps, but other types that have been far better studied — can significantly improve people's functioning.

One of the most incredible demonstrations of brain training's effectiveness was reported in the summer of 2016. With funding from the National Institute of Aging, the Advanced Cognitive Training for Independent and Vital Elderly (ACTIVE) study recruited 2,832 healthy older adults with an average age of 73.6 years at the beginning of the trial. The researchers then randomly divided them into four groups. One group received no brain training at all; two groups were taught tricks for improving memory and reasoning; the fourth and final group spent 10 hours playing a video game designed to improve their so-called "speed of processing."

Five years later, the speed-of-processing group had had half as many car accidents as people in the other groups.

Ten years later, those who had completed the most hours of training in the speed-of-processing group had their risk of developing dementia nearly cut in *half*—a finding that no drug or any other treatment has ever come close to achieving.

So what's this speed-of-processing training? Developed by a company called BrainHQ, it involves looking at a center target on a computer screen while tiny icons appear briefly on the screen's periphery. The challenge is to keep your eyes firmly fixed on the center yet still correctly identify exactly where those icons appeared. The better you get, the faster the icons on the screen's edge appear and disappear.

I don't usually like to recommend a commercial product, but BrainHQ is one of the best-researched programs available for brain training. If you want to try computerized training, I don't know of a better site to try.

Older adults are hardly the only ones who benefit from brain training. Because I specialize in surgery on brain cancer, I have long been concerned about the cognitive effects of the chemotherapy and radiation my colleagues usually offer after the operation. So-called "chemo brain" is not just a feeling of exhaustion; children, in particular, are known to experience lifelong decreases in IQ after brain surgery followed by chemotherapy or radiation. Yet pilot studies of a brain-training program called Cogmed have found that it may help to prevent or reverse such changes in children.

Offered only by psychologists who have been trained by the company, Cogmed includes a series of computerized exercises that demand close attention and focus. 3D Grid, for instance, challenges you to flick on a series of panels in the same sequence as they are briefly lit up. Another exercise asks you to type in a series of numbers after you hear them spoken aloud, but you need to list them in reverse order. Easy at first, the exercises get harder — and the results on attention and focus get better — as the sequences get longer.

Even for healthy young adults, brain training appears to pay off. One of the most rigorously designed studies in the field, published in 2018, was carried out by researchers at Oxford, Harvard, and a division of Honeywell Aerospace. After recruiting 113 students from leading universities, they tested the effects of a brain game called Robot Factory either alone or in combination with a form of mild, external brain stimulation called transcranial direct current stimulation (tDCS). After just three weeks, they found that the students who had undergone both the stimulation and the brain training saw significant gains on intelligence tests, whereas the other students did not.

BEYOND INTELLIGENCE

Memory and computational prowess are important, but if you want to be anything other than a mathematician, you will probably require a few other brain traits:

EMOTIONAL INTELLIGENCE. From the sandbox to the corner office, the ability to play well with others is huge. As science journalist Daniel Goleman showed in his bestseller of the same name, emotional intelligence is the ability to "rein in emotional impulse; to read another's innermost feelings; to handle relationships smoothly."

As ethereal and slippery as these qualities might seem, they have their basis in the brain — primarily in the frontal lobe. Although popularly characterized as the headquarters of human IQ, the frontal lobe is also where our emotional and social self-control emerges. Damage the frontal lobe, and you become an emotional wreck. Individuals affected by frontotemporal dementia likewise lose control of their emotions; they cry at the drop of a hat, laugh during a funeral, or fly into a rage over nothing.

Important as emotional intelligence is, however, we all know people who have managed to succeed without it. Many creative artists and even business leaders like Steve Jobs have been famous for their abusive tempers, their demeaning treatment of associates, and their bouts of disabling depression. So what if you're an emotional wreck? How else might you succeed?

GRIT AND DETERMINATION. Psychologist Angela Duckworth, winner of a MacArthur "genius" grant, has popularized the idea that diligence and perseverance play a far bigger role in success than do just about any other qualities. The smartest person in the room, she argues, will never do as well as the one who works the hardest. The student who keeps doing her homework, the scientist who never gives up, are the ones who go the furthest. Between the lazy genius and the indefatigable wonk, Duckworth says, bet on the latter every time. One of the few studies to look for the neural basis of grit in the brain identified a tiny region in the right prefrontal cortex, which other studies have found to be involved in self-regulation, planning, goal setting, and thinking about how past failures could have been turned to successes.

That's all good and fine. Who can argue with the value of hard work and determination? But do you really want a talentless grind to paint a masterpiece, unravel the mysteries of outer space — or perform surgery on your brain? Isn't there a role for innate brilliance?

PRACTICE, PRACTICE, PRACTICE. Actually, according to psychologist K. Anders Ericsson, there is no such thing as innate brilliance. Genius, he asserts, is simply the result of years of hard work and deliberate practice.

To support this view, Ericsson has published studies showing that a person with ordinary memory can learn to have a super-memory for numbers. An average college student, he proved, could learn to remember up to ninety random digits at a shot simply by practicing such feats of memory for months. But the catch, Ericsson discovered, was that the student would be no better at remembering words — or anything other than a string of random numbers — than when he or she began. The only talent that improved was the specific skill they practiced. What's more, Ericsson claimed, practice is likewise the key for chess grandmasters and professional violinists. Beyond a certain minimal threshold, talent or general intelligence simply does not matter.

In his book *Outliers,* Malcolm Gladwell popularized Ericsson's research by putting forward the notion of a "10,000 Hour Rule." According to this so-called rule, all you need to do in order to excel at something is to deliberately practice chess or guitar playing — or whatever — for 10,000 hours. Really? What if you did only 9,738 hours?

This, of course, is nonsense. Sure, practice improves everybody's skills, and it's absolutely essential in some fields. But are gold medals at the Olympics handed out simply on the basis of how long the athletes practiced? Will every writer who types away for ten years be rewarded with a Pulitzer? No. There are surgeons who have done 10,000 operations, let alone hours, and they remain mediocre. Talent is an undeniable component.

My view is simple: There are as many paths to success (and failure) as there are human beings. The smarter you are, the better your chances. The more emotionally balanced, the better. The grittier your determination to overcome obstacles and the longer you practice, the better you'll do. And even if the frontal lobe or a kernel of the right prefrontal cortex plays essential roles in enhancing these abilities, the bottom line is that to achieve maximum results, the entire brain must work together as a harmonious, integrated whole.

NEURO GYM: THE POWER OF SELF-TESTING

Nobody has an instruction booklet for how to win a Nobel Prize, but it's well established how to memorize any set of material faster and better. Doing so might not make you smarter in the long run, but it can definitely help you learn in the shortest time possible.

Let's say you have to learn a list of vocabulary words in a foreign language, the names of every muscle in the human body, or the lineage of ancient Egyptian pharaohs. How do you go about studying?

If you're like most people, you read and reread the material you are trying to learn or perhaps make a list or outline and study that. Practice makes perfect, right?

Wrong. Studying the same material over and over again is far less effective at improving memory than self-testing, according to studies by psychologists Henry L. Roediger III and Jeffrey D. Karpicke of Washington University in St. Louis. They have shown that after a single review of material, repeated self-testing enhances learning far better than does repeated studying. So test yourself and find the edge of your knowledge: that's where learning happens.

Let's test you right now on how many facts you remember from the section above on brain training:

1. How much was Lumosity fined by the FTC?
2. How many accidents did the people who did speed-of-processing training have compared to those who did not?
3. What does tDCS stand for?
4. How many years after speed-of-processing training was conducted did researchers see an effect on participants' risk of developing dementia?

After you have checked your answers, go ahead and read the next chapter. Then come back and try to answer these four questions again. I bet you do well. Self-testing is a powerful tool for helping you remember!

3
THE SEAT OF LANGUAGE

I t all started, Marina told me, with a simple word: *pen*. Six months earlier, the thirty-three-year-old English teacher had handed out a quiz to her class of high school students when one of them said that he had nothing to write with.

"Here," she told him, "take my—" and that was when she realized that she couldn't produce the word for the writing implement she held in her hand.

Over the next five months, other common words began failing her. The lapses grew more frequent, but she kept trying to dismiss them as glitches. Secretly, she was increasingly gripped by worry because her mind's intentions weren't finding voice.

Marina had been born and raised in Chile until the age of twelve, when her parents immigrated to southern California. Spanish was her native language, and English was her second. And now, as English words became foreign, she had begun substituting their Spanish equivalents to keep her speech fluid. "Pen" was undiscoverable, but "pluma" slid naturally into its place. Her original tongue had resurfaced as her lexical life vest.

After all, who hasn't been frustrated by the occasional word or name stuck on the tip of their tongue? But at her age, to have this happen over and over again concerned her family doctor. He ordered a scan, which showed a small amorphous dark patch in the center of the left temporal lobe. The radiology report mentioned "suspicious for malignancy."

Had it shown up as a bright white splotch on the MRI, that would have meant it was more likely a dangerous glioblastoma. These high-grade, aggressive tumors send chemical signals to prompt the growth of new blood vessels to nourish them. As a result, they can suck up the contrast agent the nurse injects into the person's vein, thereby making them look bright. Whereas a gray-looking mass, like Marina's, was more likely to be slow-growing and less invasive, making it more amenable to treatment.

I said to Marina and her husband, "I have some answers to why the words are getting stuck, but I also have some uncertainty to discuss." I showed them the key brain images and pointed out the shadowy appearance of the tumor. Cancer? Yes, but a rare type of brain cancer that can be surgically cured if removed completely. They were blanketed with despair, but hearing the word *cure* allowed for a hint of hope.

Brain surgery to remove this tumor would be no easy task, however, because Marina's tumor was located in her left frontotemporal region: the seat of language. Cancers take on a myriad of three-dimensional shapes, and no two are alike. It's a different enemy every time, and hers had taken refuge behind exquisitely eloquent brain tissue.

Had it been elsewhere, I would have abundant safe zones through the brain's surface to access deep-seated tumors. But in her, my surgical approach would have to be through this inscrutable "seat of language." To get it all, I would have to find windows of safe entry in the perilous region that rests on the banks of the deep sylvian fissure, separating the left frontal lobe from the left temporal lobe, which is marbled throughout with neurons critical to language. Injure the wrong section, and you can lose the ability not only to speak but to understand signs and gestures. Marina was at risk of losing communication at its most fundamental level.

BROCA AND WERNICKE

Allow me to step back, for a moment, from Marina's story to reveal a bit more about the location of language in the brain. Scientists vigorously debated the matter in the nineteenth century. Some insisted that it was nowhere and everywhere, that you could remove any section of the brain and it would not entirely wipe out the ability to speak and understand. The first piece of evidence that language has a particular home address in the brain came from a French shoemaker who, in 1840, at the age of thirty, lost the ability to speak — except for one word: *tan*. Louis Victor Leborgne could understand what other people said; he could reason; but until the day he died, *tan* remained the only word he could say or write. Asked a question, he would typically say it twice: "Tan tan." Admitted to the Bicêtre Hospital, a psychiatric asylum just outside of Paris, he soon gained a nickname: Tan. Over the course of the next twenty-one years, he became paralyzed on his right side and developed gangrene.

Days before his death in April 1861, Tan was visited by a physician with a special interest in speech: Pierre Paul Broca. When Tan died, Broca conducted an autopsy on his brain, finding an area of dead tissue, likely due to syphilis, at the back and lower portion of the frontal lobe, near the fissure separating it from the temporal lobe.

A few months later, Broca met another patient at the same hospital with nearly the same peculiar disorder. Lazare Lelong, eighty-four, could say only five words: *oui* ("yes"), *non* ("no"), *trois* ("three"), *toujours* ("always"), and *Lelo* (a garbled version of his own name). When Lelong died, Broca autopsied his brain, too, and found an area of dead brain tissue in almost the exact same spot as Tan's.

Now known as Broca's area, this small region of the brain is recognized as critical to the production of speech. But another area, discovered soon after by the German neurologist Carl Wernicke, underlies the ability to *comprehend* speech. A person who suffers an injury to Wernicke's area — near the same sylvian fissure but on the temporal side — will continue to speak fluently but in a meaningless word salad.

For a hundred years after Broca and Wernicke revealed their findings, scientists thought that these regions were precisely where the two doctors said they were. By the time I began my training, however, it had become clear that these were only approximations. Language, it turns out, has a fuzzy address.

NEURO GEEK: THE NEUROSCIENCE OF BILINGUALISM

Marina's bilingualism was actually a gift to her brain. People who learn a second language gain significant benefits in cognitive health that last a lifetime. That shouldn't be surprising because, as brain-mapping shows, different areas of the brain handle different languages. Busy neurons are thriving neurons; those without any assigned tasks tend to wither.

How exactly does bilingualism pay off in brain performance?

IMPROVED ATTENTION. British researchers at the University of Birmingham recently recruited 99 volunteers, 51 of whom spoke only English, the remainder being bilingual in English and Mandarin since childhood. The English-only speakers performed slower on two out of three tests of attention. Being able to switch between two languages, the researchers concluded, improves a person's ability to maintain focus and attention.

Dozens of other studies have demonstrated similar benefits in attention and focus for people who speak two languages. The benefits are due in part to the fact that a bilingual person's brain must actively suppress one language when speaking another. Being able to handle that extra workload results in stronger overall control of attention. Imaging studies have shown benefits in both the prefrontal cortex and subcortical regions. The better a person speaks a second language and the earlier it is acquired, the more gray matter has been seen in the left parietal lobe cortex. More white matter has likewise been seen in both children and adults who speak two languages.

IMPROVED LEARNING. A four-year study of children in public schools in Portland, Oregon, randomly assigned some of the kids to usual English-only classrooms and some to dual-language classrooms where they learned Spanish, Japanese, or Mandarin. By the end of middle school, the dual-language kids had gained a full year in English-reading skills on their peers.

In another study, English-speaking children placed in a Spanish-immersion program performed better on tests of working memory and word learning than did children who remained in an English-only program.

PROTECTION AGAINST DEMENTIA. An extraordinary study published in 2007 by researchers in Toronto showed that people who spoke more than one language developed symptoms of dementia about four years later than people who spoke only one. Since then, other studies in Montreal, India, and Belgium confirmed the protective effect against dementia in bilinguals. As a recent review in the journal *Current Opinion in Neurology* concluded: "Lifelong bilingualism represents a powerful cognitive reserve delaying the onset of dementia."

The message is clear: If you have kids and know a second language, using it around them will build their brain power and cognitive reserve.

BRAIN MAPPING

So if I was going to remove Marina's tumor without destroying her ability to speak and understand language, I needed to map the surface of her unique cortical canopy, exploring for safe versus language-sensitive spots, while she was awake and able to guide me. I needed to become a cerebral cartographer in search of tiny islands of brain tissue that could be parted and serve as portals to dive deep inside her brain.

I told Marina and her husband that I would have to remove her tumor while she was awake. She would need to tell me, by her ability to speak — or not — where it was safe for me to begin dissecting. And because she was bilingual, I would have to check each spot twice: once for English and again for Spanish.

Three weeks after our meeting, I stood beside Marina as her eyes fluttered open. We had to put her to sleep for the beginning of the surgery, when her scalp, skull, and dura were opened. It would have been too painful for her to be awake for that portion. But because the brain itself has no nerves to feel pain, the anesthesiologist was now able to dial back the sedative in preparation for the real work.

"Welcome back," I said to Marina. "You okay?"

"Groggy. Is it . . . open?"

"It is. . . . Let me know when you're ready," I said.

Days prior, we had gone over the operation and her role.

In my left hand, I held the electric stimulator, a device the shape and size of a fountain pen. The twin tips are forked like a snake's tongue and release a miniscule electrical current that runs from one to the other. The current works like a taser to the neurons between its prongs. It allows me to temporarily stun a tiny piece of tissue. The brain cannot feel my touch; it does not have the ability to know when it's being touched, cut, or manipulated. But the zapped neurons are momentarily stunned, resulting in the loss of whatever function they possess.

The neurophysiologist asked her to count to ten. Marina did. He asked her to sing the alphabet. She complied. Now he repeated both requests in Spanish, her native tongue. She performed flawlessly. We were ready.

I lowered the stimulator onto a spot on the edge of Wernicke's area. The neurophysiologist posed Marina a list of questions in English and Spanish, each of which she answered flawlessly. He showed her objects and asked her to identify them. All went well.

This spot, I inferred, is safe. No harm to her language abilities would result from my inserting a scalpel there, if I must. To mark it, I placed a tiny white square of confetti-like paper directly onto the brain. The slippery surface held the dry paper with no need for adhesive.

I moved to an adjoining area. The neurophysiologist had Marina singing the alphabet when I zapped the spot with the stimulator. Marina kept right on singing. Then she went through the song again in Spanish. I waited until she was halfway through to zap it. She stopped at N as if I had pressed a mute button. Her speech arrested.

This spot was a no-go. I placed a red piece of confetti on it with a letter S for Spanish.

An hour later, the glistening topography of the language-critical area of her left brain was covered in a mosaic of white and red confetti. Some of the reds had the letter E for English, some had S for Spanish, and some had E/S for both. White was "takeable," meaning I could dive through the brain's surface underneath the white pieces of paper to reach the tumor below without robbing Marina of speech.

With the testing portion behind us, I began to perform the surgery. I removed a white-marked area and carved a one-eighth-inch corridor of tissue with my scalpel. Because the tunnel was so small, I used a surgical navigation system to see below the surface. It combined live 3-D images of Marina's brain to show me exactly where I was going.

Two inches deep, I reached one edge of the tumor. I switched to suction and took out as much from this angle as possible. But because the suction tube is rigid and my point of entry on the surface was surrounded by red no-go tissue, I couldn't twist or toggle my instrument to get all of the tumor. I had to start again from another white-flagged point on the surface to come at the tumor from another angle. Carefully working the depth and incessantly attentive to the superficial portals, I got after it.

Throughout the procedure, the neurophysiologist had Marina talking and singing as I dissected through portals that made the brain's surface look like Swiss cheese to reach all the edges of her cancer. Whenever she stopped, I stopped and backed away my scalpel from a language-precious location. We were cosurgeons, Marina and I. Her voice drove me; her silence made me brake.

Three hours later, I closed the last stitch on her scalp to conceal what occurred inside, but our work was not yet done. Despite the reassurances of her speech fluency during surgery, I would not be settled until I met her in recovery, to be sure she could understand me and talk to me in both English and Spanish.

Outside the OR, her family waited and read my body language as I approached. I wasted no time and immediately told them the essentials. She's alive, moving, talking. Those four words put them at ease, and then we spoke in greater depth.

The day after surgery, her MRI report from the neuroradiologist said "gross total resection," or GTR. My review of the scan showed the same; as best we could tell, every last bit of the tumor had been removed.

Every three months for the next year, she returned for another brain scan, each one showing an immaculate inspection, with no evidence of tumor. Her speech improved. She was teaching again. Worry and fear receded.

A DREADFUL CHOICE

On the fifth follow-up visit, a year and three months after the operation, a small grape-sized shadow appeared on Marina's brain scan. Despite the MRI showing no trace immediately after surgery, she had now developed a tumor recurrence. Malignant seeds that are invisible to scans and scalpels persist, despite our best efforts, and become the cell of origin for a second tumor that is sometimes biologically more aggressive than its ancestor.

I showed her the scans. She took it hard. I would, too. Not only was the next step another awake brain surgery, but the recurrent tumor might reveal itself to be more aggressive than before, taking away the chance for a cure.

Back to the operating room. The monitors on the wall showed me the photographs I had taken with red and white confetti revealing her original language mapping: an archipelago of language function with some islands of brain tissue dispensable and others vital. But the original map could no longer be trusted. Her language areas had undoubtedly reorganized in response to my dissection.

The result could pose a dreadful choice for Marina: This time, I might have to cut through a language-critical area. But which language?

I used a biopsy needle to sample the tumor and sent it to pathology, down in the hospital's basement, for a quick determination of its grade. For thirty long minutes, we waited to hear the result. Finally, the pathologist called, and the nurse held the phone to my ear. It was still a Grade 2! That meant a second attempt at a GTR could still be curative.

We repeated the mapping. English now inhabited a tuft of brain tissue that originally housed Spanish. Cortex that led to no speech arrest previously had now become critical for fluency. The areas I could safely flag with white were fewer. And keep in mind, one false move into a spot no wider than one-eighth inch could remove an entire language.

My hope was that the original dissection would grant me the angle to remove the tumor, but sometimes new windows need to be made for access. For Marina, an English teacher, the choice was nearly impossible. Was she willing to lose her ability to speak English if it meant removing all the tumor? Since Spanish was her native tongue, if the brain region controlling Spanish was injured, *both* languages might be forever out of her reach. That was something neither she nor I would choose, the complication to be absolutely avoided.

There actually was a second approach: I could leave behind any part of the tumor not reachable from the safe spots on her brain's surface. Then she could still get chemo and radiation and hope for the best. It would not be a cure, just a temporary pause in her tumor's growth. But she would awaken from the surgery with her linguistic abilities fully intact.

A week earlier, Marina, her husband, and I had discussed the two scenarios if the tumor hid beneath eloquent cortex.

"English," she said. "Take English if you have to. I need at least twelve years."

"Why twelve?" I asked. "We're hoping for a cure."

"That's when my youngest would be in college in case I got sick," she replied.

And that's what I did. To completely remove her cancer, I dissected through brain tissue that a little over a year ago had been irrelevant to her speech but that now mysteriously housed English. At that moment she was no longer bilingual.

For the last five years, her scans have remained negative. She is cancer-free.

We speak in Spanish.

NEURO GYM: HARNESS THE POWER OF A SECOND LANGUAGE

If you already know a second (or third) language from taking foreign language classes in school, count yourself lucky. If you grew up bilingual, hearing more than one language early in life, count yourself even luckier. But what do you do if you want the cognitive reserve of a backup language but haven't yet learned one?

I recommend that you take a class that requires you to show up in person. There are plenty of language programs offered online, of course, and they make a lot of claims for how quickly you can learn. Some of them are cheaper than in-person classes. But you get what you pay for, and nothing beats a real classroom setting. It's highly motivating to lay out cash and commit to a series of scheduled classes. The fact that your teacher and fellow students will all notice if you haven't yet learned how to say "Where is the bathroom" has a real effect on the likelihood that you will study between classes. More importantly, student-to-student conversations work wonders at improving language skills.

If you still prefer to use an app, *PC Magazine* reviewed a bunch of them and concluded: "The best free language-learning app is Duolingo, hands down." The *New York Times* also recently recommended Duolingo, and the *Wall Street Journal* called it "far and away the best free language-learning app — and a good deal better than some of the pricey ones." Either way, whether you dip your feet into a new language by taking a class, downloading an app, or traveling, the exercise will serve your brain well.

4

UNLEASH CREATIVITY

H e had been sitting calmly on his gurney in the hallway of the emergency room for hours, the nurse told me, saying nothing, moving little, dressed in greasy rags, looking neither frustrated nor exactly calm. This being an East L.A. hospital, I quickly read that he was a homeless person.

"Meet the famous TV producer," the nurse said with a wicked grin, out of his earshot.

"TV producer?" I asked, assuming she was joking.

"That's what he always says," she answered. "The famous homeless TV producer."

I checked his file. William, it said, had been in and out of shelters and was a semiregular visitor to the hospital's ER, usually with mental health issues. Tall, gaunt, with clumpy, thick, golden hair and beard, he looked like an emaciated lion. I walked over and introduced myself and asked him what happened.

"My head," he said without emotion. "Hurt it."

I asked if I could check out the back of his head. Looking back at me with vacant eyes, he said yes.

He had a gash in his scalp, with a glint of ivory-colored skull visible underneath.

"How'd this happen, William?"

"I fell."

His arms had no tracks from injections. No rigs for shooting up were in his clothes. According to tests that had been done when he arrived at the hospital, no drugs were in his urine and no medicines or drugs were in his blood.

"The nurse says you were a TV producer?" I asked.

"Yes," he said. He mentioned two sitcoms and an hour-long cop show that had been popular in the eighties.

"You produced those?"

"Actually, I was the showrunner," he said in a robotic cadence.

"So what happened, how'd you end up homeless?"

"Got fired."

Looking down at his file, I saw that he was single.

"You had a wife?"

"Divorced."

"Kids?"

"Two."

"You drink, William?"

"No."

"You get high?"

"No."

Since there was no skull fracture, I told the ER doctor to clean and staple it shut. Still, the clinical pieces didn't fit — what could explain this former television showrunner's flat demeanor? So I ordered an MRI of his brain. MRIs don't expose patients to radiation, and perhaps I would find a "zebra," medical slang for a rare diagnosis.

Five hours later, coming out of surgery on another patient, I took a call from the radiologist.

"Got something for you," she said.

I walked down to her office, and she brought up the MRI on her computer. The images popped up slice by slice. First I could see his nose, then his eyes, then his forehead bone, then the tips of the frontal lobes just behind the forehead, and then a bright, white sphere a little larger than the size of a chicken egg, nested between the right and left frontal lobes.

You didn't need a medical degree to see it or know that it needed to come out.

Planum sphenoidale meningioma: a big name for a rare, slow-growing tumor right smack in the middle of the frontal lobes, the seat of our most advanced intellectual and executive-level skills. But it doesn't eat into the surrounding brain tissue, so it's not considered cancerous per se. Likely it had started its glacial growth decades ago.

When it had first begun expanding, the frontal lobes had been accommodating, moving out of its way. But at some point, as they were spread so far apart from the midline, they were backed against the left and right insides of the skull. The gradual nature of this gentle, constant pressure allowed the lobes to shape-shift. Just as water over millennia can shape stone, over years the brain can be molded by steady and persistent pressure. But the symptoms — extinguishment of emotion and motivation, loss of creativity and self-control — are often confused with major depression or dementia, particularly because they come on so slowly as the tumor grows and squeezes the frontal lobes harder and harder against the unforgiving skull. Due to the slow onset and resemblance to other issues, we routinely fail to diagnose it early.

Surgical removal was the only treatment, but the surgery would be no easy task. A few days later, I was folding William's scalp off his forehead, allowing it to flap down over his eyes and nose. I made multiple holes and, using a small fine chisel, lifted off the portion of his skull covering his forehead. In protest, the last thin, bony connections cracked and fractured with a sound like splintering wood.

Welcome to Mother Nature's masterpiece: the ultimate marble for a surgical sculptor. The midline is the most dangerous part. A tense, gray tissue runs front to back in the middle of the skull, like a keel dividing the hemispheres, but it also houses a giant vein that drains blood flowing out from the brain tissue. I tied it off, cut off a few inches, and beheld my target: William's two frontal lobes and the tumor growing between them.

Over the next few hours, I cored out the ball of tissue — it had the texture of overcooked scallops — until it was hollow. Millimeter by millimeter, I peeled it off its embrace with the brain tissue. The goal was to never violate the surface of the brain while dissecting off the tumor.

When I was done, the two frontal lobes were still stuck in their distorted shape, accommodating a meningioma that was no longer there. But I knew that over the coming months, they would relax back into their normal voluptuousness.

I also knew that this would take time. I saw no noticeable change in William's Spock-like, emotionless demeanor when we discharged him to the care of a rehabilitation facility. But two weeks after that, when he came in for his first post-op checkup, I saw him smirk for the first time. A few months later, I learned he had a job, a car, and an apartment. He may never get back his job as a television producer, but he had reconnected with his two adult children, who had once thought that their dad was a drug addict.

Thinking back on William's case reminded me how dependent we all are on the delicate architecture of our brains. The brain gives, and the brain takes away. We like to think of our most advanced abilities — our creativity, our intelligence — as something that defines us, something that we personally brought into being. Nobody says that a person "has" creativity; we say he or she "is" creative. But let a tumor grow between our frontal lobes, and we learn — surprise! — those great gifts were simply on loan. Amazingly, disrupting one frontal lobe can leave people functioning normally. It's what *both* frontal lobes do in concert that sprouts our highest cognitive function: creativity.

So what does a story about a TV showrunner's brain tumor tell us about the nature of creativity, and how did that tumor pressing against William's frontal lobes make him lose his? Where does the creative spark come from?

NEURO BUSTED: THE LEFT BRAIN/RIGHT BRAIN MYTH

One of the most ridiculous ideas out there about the brain's role in creativity is the "left brain/right brain" myth. It all began in 1973, when the *New York Times Magazine* published an article about the research of Nobel Prize–winning researcher Roger W. Sperry. "We Are Left-Brained or Right-Brained," the article asserted. Supposedly, the story went, the right side of the brain is the creative or artistic side, the left brain is the logical, analytical side, and each of us tends to favor one or the other.

It sounded great, and the idea promptly became something that everybody "knows," but it had one problem: it's wrong and has since been demolished by decades of research. What *is* true is that parts of the left hemisphere are intimately involved in spoken language and in mathematical tasks such as counting or remembering your times table. But the notion that there are "right-brained" people who are more creative and "left-brained" people who are more logical is simply not true. The most definitive annihilation of this idea came in a 2013 study published by researchers from the University of Utah. They reviewed MRI brain scans of more than a thousand people between the ages of seven and twenty-nine to see if they could find any support for the theory that some people use their left brain more while others use their right brain more.

"Our data are not consistent with a whole-brain phenotype of greater 'left-brained' or greater 'right-brained' network strength across individuals," they concluded. In other words: math geeks and computer programmers use both sides of their brain equally, as do painters and poets.

LITTLE FIRES EVERYWHERE

There's no question that the frontal lobes, the part of your brain pressing up against your forehead, are essential for any creative work. As the most advanced part of the brain, they keep us organized, motivated, and on-purpose in ways that nonhuman animals simply can't fathom.

But they can't do it alone.

Consider the tiny cerebellum, the bundle of brain cells lying below and behind the main body of the brain, jutting out from the brain stem like a mushroom on an old tree. When I went to medical school, I was taught that the cerebellum's only job was to coordinate fine muscle movements learned from years of practice. But new studies published in the past few years have shown that increased activity in the cerebellum is directly linked to creative problem-solving. It coordinates creative thinking, we now believe, just as it coordinates the fine muscle movements of an athlete.

But it takes the *whole* brain, working and communicating together, harmoniously and in sync — like a symphony orchestra or a football team — for creativity to happen. The spark comes not from any one point but from all the points uniting in a grand network.

How do neuroscientists know when different parts of the brain are communicating with each other? We eavesdrop with functional MRIs, which take 3-D movies rather than still photos of the working brain. This allows us to localize which parts of the brain are more or less active from second to second, as evidenced by how much blood is drawn to any given area. (Yes, your brain cells use more blood when they're hard at work, just like your muscle cells do when you're running.) With the help of sophisticated computer programs, we can calculate how likely it is that activity in a neuron over *here* is in sync with the activity of a neuron over *there*. They do that for tens of thousands of neurons simultaneously, allowing them to see if, in a manner of speaking, the lightning bugs in your yard are all blinking in coordination or if they're all random and out of sync. Creativity, we now understand, requires brain cells to light up in coordination.

Everyone has a deep well of creativity within themselves just waiting to be tapped. Surprisingly, along with the well-known deterioration of memory and disruption of mood that come with dementia, some people with Alzheimer's disease develop newly discovered artistic abilities. The gradual release of their "hidden" artistic talent is most often exhibited through dramatic accentuation of the ability to draw or paint. This is both well described in the medical literature and the subject of an excellent documentary, *I Remember Better When I Paint*. In a similar vein, hidden savants have demonstrated elite musical or mathematical ability, seemingly instantaneously after car accidents or, in one case, after being struck by lightning. These extreme examples suggest the creative potential lurking inside each of us.

Here is how I have found my own key to generating creative insights.

MY METHOD FOR CREATIVITY

Francis Harry Compton Crick, one of the four scientists who were awarded the Nobel Prize for figuring out the structure of DNA, "retired" to San Diego to explore what he felt was the greatest remaining question for science: the origin of human consciousness. I had the opportunity to meet him in La Jolla, California, the once-sleepy town in northern San Diego County at the very bottom of southern California that boasts a beautiful plateau overlooking the ocean. There, atop the cliffs, stand four of the most prominent biological research institutes in the country: Salk, Scripps, Sanford-Burnham, and UCSD. Reported to have the highest per-capita rate of neuroscientists in the world, La Jolla is where I performed my PhD work and ultimately defended my thesis.

In a brief conversation, Crick left an indelible impression on me with his advice: "Nerds make good technicians; scientists need to be creative." A decade later, those words increasingly resonate.

I now know that the hardest part about running my laboratory at City of Hope is coming up with original ideas about how the brain works and about how cancer exploits the brain's inner workings. It isn't easy to have a fresh take on Mother Nature. One can't simply rely on data mining or crowd-sourcing to divine new insight. The exciting part of this challenge, however, is the recognition that one individual can, at a stroke, outdo literally dozens of teams of scientists with a single, clever idea diligently explored and demonstrated to be true or not true.

My method for generating new ideas for my research is an extension of a longstanding habit of mine in surgical planning. The night before particularly challenging operations, I always carefully review the images of the patient's brain and brain tumor just before going to bed. Then, while falling asleep, I imagine rotating the tumor, seeing the surrounding, dangerous terrain that I must avoid or traverse. Upon waking the next morning, I again take a few minutes to revisit the shapes and contours. This simple practice, I have found, helps to deeply imprint on my mind a spatial awareness of the anatomy that I need to dissect through and around.

As an extension of that approach, two nights of every week, just before going to sleep, I read articles that are directly or tangentially related to any experiment I'm struggling with. In this way, by rolling others' work in my mind as I grapple with my own challenges, I make new connections between what is already known and my laboratory's provocative findings.

It's not only for scientists like myself that the borderlands between slumber and wakefulness have proved so fertile. As Salvador Dali wrote in *Fifty Secrets of Magic Craftsmanship*: "You must resolve the problem of 'sleeping without sleeping,' which is the essence of the dialectics of the dream, since it is a repose which walks in equilibrium on the taut and invisible wire which separates sleeping from waking."

Since we are all Dalis of creativity when we dream, it may be that the time when one drifts off to sleep (*hypnagogic* is the neuroscientific term) and the time when one is partly asleep and transitioning to the awake state (*hypnopompic*) may offer the brief portals when subconscious creativity can be accessed for creative insights.

This blending of sleep and awake states can actually be detected with an electroencephalography (EEG). On EEG tracing, these periods show *both* the alpha waves of being awake and relaxed *and* the theta waves of the sleep state. This is the only time we know of when alpha and theta waves overlap. So, focused awareness during the dusk

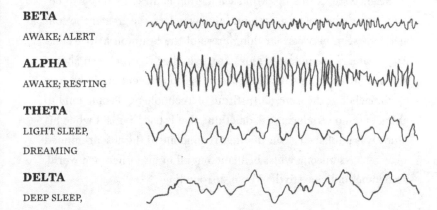

BETA
AWAKE; ALERT

ALPHA
AWAKE; RESTING

THETA
LIGHT SLEEP,
DREAMING

DELTA
DEEP SLEEP,
DREAMLESS

and dawn of your sleep could hold the key that unlocks the creative potential of your subconscious.

But there are plenty of other ways to jack up your creativity. We have untapped potential.

NEURO GYM #1: LET YOUR MIND WANDER

Every great advance in music, biology, and astronomy, in literature and technology has involved overturning orthodoxy and doing what the experts said couldn't or shouldn't be done. Rule-followers, on the other hand, are like good drivers: they keep their thoughts in their own lanes, never jumping the curb. But their focus has a dark side: it inhibits the random, out-of-left-field connections that define true creativity.

The brain, you see, is *not* a computer, despite oft-repeated claims to the contrary. The brain is a living thing, much more like an overgrown garden than an orderly filing cabinet. And mind-wandering through your own garden of thoughts, memories, feelings, and desires is a sure way to discover your inner creative self.

Science backs this up. Mind wandering is directly linked to enhanced creativity. The more your mind wanders, the greater the connections seen between far-flung areas of the brain on MRI exams. Daydreamers are not only more creative, they've even been shown to be smarter on certain tests, according to a recent study by Eric Schumacher at the Georgia Institute of Technology. "People tend to think of mind wandering as something that is bad," he said when his study was published. "Our findings remind me of the absent-minded professor — someone who's brilliant, but off in his or her own world, sometimes oblivious to their own surroundings."

I don't mean that students should stop paying attention to their teachers or that you should spend your time at work staring into space. Of course, you must pay attention long enough to build a knowledge base. But creativity requires a *balance* between homing in and spacing out, between mastering material and going off on a tangent. Darwin did a deep dive into biology and botany, as well as anyone, before he turned it on its head with evolution; and you, too, need to grind out the hard work of learning what's known before you can discover what was never before imagined.

NEURO GYM #2: GET PLAYFUL

Creativity is a kind of grown-up version of play, so perhaps it isn't surprising that childhood play and adult creativity are intimately linked. Even for engineers, architects, and scientists, early experiences in unstructured free play — especially pretend or make-believe play — is a boon to later creativity.

Psychologist Sandra Russ of Case Western Reserve University has spent more than two decades studying the connection between childhood play and adult creativity. Through play, she says, "Children learn how to process emotions and develop the cognitive processes that help them experiment with the lifelong skill of problem-solving." She encourages parents to allow their children to play with ordinary boxes, pans, even furniture. "Play teaches them some lifelong skills that they can take into adulthood and use in creative ways," she says.

And by play, Russ isn't thinking of travel soccer leagues. Structured, adult-organized activities — gymnastics lessons, chess clubs, play dates — have all but replaced ordinary free play in which kids have to figure out for themselves what to do. Are we in danger of raising a generation of well-behaved, perfectly socialized robots? No one can say for sure. But the hyperstructured way in which many children are raised now is not the way most kids grew up prior to the twenty-first century.

That's why I also don't buy into the endless tests, and test prep, that many children in elementary schools these days are forced to endure. All that rigor can kill their imaginative spark. Incredibly, some schools have even eliminated the recess period — my sons would revolt! Other schools, however, are going in the opposite direction: they're adding extra recess time to the school day while eliminating all homework in elementary school.

I'm not saying you should take their cell phones and tablets away or cancel their play dates and ballet classes. But this is what I know: unstructured free play in childhood is a foundation of adult creativity. Personally, I try to give my three sons some space to explore, invent, and figure things out for themselves — even take some risks. Schools in Britain, in fact, have begun intentionally "bringing in risk" to their playgrounds and classrooms, allowing children (under supervision, of course) to play with saws, scissors, old pieces of wood, and even fire. Allowing children the opportunity to go a bit rogue is not a guarantee, but it's the best elixir for them to mature into multidimensional adults.

For instance, on a recent family vacation, we put our thirteen-year-old in charge of navigating our way through the Tube (the London subway). He picked the correct train but the wrong direction. We let him figure it out, and he did. Creativity, after all, requires the confidence to know that mistakes happen and are part of the process. Fear of failure keeps too many people from daring to express themselves.

NEURO GYM #3: GET OUTSIDE

Sure, hipster havens in Brooklyn or the Bay area with open work spaces are famous incubators of innovation, but nature also has a place in nurturing creativity.

David Strayer, a psychologist at the University of Utah, enrolled thirty men and twenty-six women in a simple experiment. Half took a test of creativity the morning *before* a four- to six-day backpacking trip; the other half took the test afterward. Result: Those who took the test *after* their wilderness experience scored 50 percent higher for creativity than those who took it before the trip.

Nature, of course, has been celebrated as a tonic and inspiration since pretty much forever. But consider: U.S. children now spend less than half an hour per day engaged in outdoor play or sports compared to nearly eight hours a day watching TV or (increasingly) swiping away at their cell phones. Meanwhile, the average number of trips per person to national parks and forests has declined by 25 percent over the last thirty years.

But you don't have to spend a week hiking in the forest to have your creativity nourished by the outdoors. Even a half-hour walk near your home, office, or school will do. Einstein made a habit of walking the mile and a half to and from his office at Princeton University every day. A little exercise, some fresh air, the passing of the seasons: it's all fuel for your creative brain.

Each of my three recommendations for boosting creativity have a common thread: I am urging you to break the routine and spend more time goofing around. Sleep, dream, play, take a walk: do anything but work. And I say this to you as someone who, remember, went through years of medical school, graduate school, and neurosurgery training. I certainly understand the need for work and focus, for studying, for spending eighteen hours a day at it. But human beings are not automatons. We are called for greater things.

NEURO GEEK: MICRODOSING FOR CREATIVITY

Recently, in the tech hubs of Silicon Valley and Silicon Beach in California, there has been a resurgence of interest in psychedelics, drugs that lead to profound alterations in perception and cognition. Administering the agents at very low doses is called microdosing. Psychedelic drugs remain illegal in the United States, unless you're part of the United Church of Christ in Kentucky, where the U.S. Supreme Court has made an exception for religious use of ayahuasca, or a cancer patient enrolled in a monitored clinical trial in which psilocybin is being administered in the hope of reducing the anxiety and depression associated with the existential crises that can accompany a cancer diagnosis.

With that said, what happens at the molecular level when you "trip out" has an interesting effect on creativity. To "trip out," you have to trip up your frontal lobes. Although creativity arises from the frontal lobes, it almost always takes a backseat to the dominant responsibility of those same lobes: the executive and planning function that aims to get through the day as efficiently as possible. So with time, mental habits take hold that reinforce efficient behavior by forming preferential electrical pathways and networks, like freeways directing the major flow between major cities. What psychedelics are thought to do is to disassemble the freeways temporarily, leaving only a dense and evenly distributed network of roads. This dissolution greatly expands the diversity of connections and, as a result, allows unexpected and original thoughts.

Using psychedelics to facilitate creative release is not new by any means. Since their discovery, they have long and notoriously been used by musicians, artists, and scientists. Yes, scientists. Although one press report called him "perhaps the weirdest human ever to win the Nobel Prize for chemistry," the person who invented the molecular technique to faithfully expand tiny specimens of DNA (called the polymerase chain reaction or PCR), Kary Mullis has been unabashed about his LSD use, saying how the idea came to him on how to unwind a double helix and make two copies at least "partly on psychedelic drugs."

Other icons of creativity have been more than open in discussing the value of psychedelic use to their lives and creative works. Steve Jobs was once quoted as saying, "Doing LSD was one of the two or three most important things I have done in my life." Along those lines, he added his opinion that Bill Gates would "be a broader guy if he had dropped acid once."

I am not suggesting that you start dropping acid. Not only is it illegal and potentially dangerous for some people, but much more research needs to be done. But if the early studies hold up, tiny amounts given in a controlled setting — not enough to cause hallucinations but the equivalent of a single sip of wine — might just help some of us become a little, um, *broader* in our creative abilities.

5

SMART DRUGS, STUPID DRUGS

I do drugs.

There, I said it.

I do mind-altering drugs. Even as I write this, I have a buzz on: a serious coffee buzz.

All thanks to caffeine, the most widely used drug in the world.

And before I tossed back my espresso, as I was settling in to write at around 2 p.m. this lazy Sunday afternoon, off duty, in January, I had a bottle of pale ale.

Alcohol: another widely used drug.

For me, it's sometimes what works when writing creatively. One beer — no more — followed by one coffee. Plus my Spotify playlist: first Sia, then Nirvana, then, who knows, some Rihanna or Linkin Park.

Strangely, I rarely drink alcohol or coffee at any other time, and I never drink coffee before going to work at the hospital. If the prospect of opening up somebody's skull isn't enough to wake you up without a cup of coffee, then maybe don't do it. Besides, peaks and valleys in energy just disrupt the surgical cadence.

But sitting down at a computer to gaze at a blank Word document?

For me, that requires fortification. Please, bring on the espresso and ale. The latter disinhibits my frontal lobe a bit, allowing me to zoom out and craft a compelling narrative. And the caffeine is like a kick to my corpus callosum, helping me deliver details and content.

Even then, I have found that I write best for only ninety minutes at a sitting, despite the fact that I have done surgeries for up to eighteen hours straight.

I wouldn't dare attempt to explain it. I am but a stranger to myself.

I share my own experience because most of us do use mind-altering drugs of one kind or another. Even in Utah, where 60 percent of residents are Mormons (forbidden by their faith to use alcohol, tobacco, coffee, or tea), prescription drug abuse is among the highest in the nation.

After all, animals have co-evolved with plants since Earth was an infant: the very breath we take is what plants exhale (oxygen), while the carbon dioxide we exhale is what they breathe in. The opioids and nicotine that grow in poppies and tobacco have their analogues in our brains, where opioid and nicotinic receptors send and receive messages of reward and pleasure, movement and alertness. Even marijuana connects with the brain's cannabinoid receptors involved in pain, appetite, mood, and memory.

These days, the talk is all about "smart" drugs to enhance memory and concentration, typically involving the use of prescription pharmaceuticals that were originally designed to treat attention-deficit hyperactivity disorder (ADHD) or sleep disorders. A recent online survey by the scientific journal *Nature* found that one in five respondents used drugs to improve their focus, concentration, or memory. Another survey of tens of thousands of people around the world found that 14 percent said they had used a stimulant (like Ritalin) sometime during 2017, compared to just 5 percent in 2015.

Lost in the conversation about "smart" drugs is that dose matters, as do the age and genetic predisposition of the person taking it. Not everything natural is good, and not every chemically synthesized drug is bad. Most important, there are smart drugs, and there are stupid drugs.

So here is my guide to help you figure out which is which.

ALCOHOL

About 88,000 people die each year in the United States due to alcohol use, according to the Centers for Disease Control and Prevention. Internationally, the burden is 3.3 million per year, accounting for nearly 6 percent of all global deaths. Even without the deaths, 15 million U.S. adults and another 623,000 adolescents have what doctors now call alcohol use disorder: a chronic, relapsing brain disease characterized by an inability to stop or control alcohol use despite adverse social, occupational, or health consequences.

Aside from the obvious cognitive symptoms of excessive alcohol use, like not remembering what you did last night, chronic overuse can cause serious brain disorders like Wernicke's encephalopathy (mental confusion, paralysis of the nerves moving the eyes, difficulty with muscle coordination) or Korsakoff's psychosis (all the fun of Wernicke's plus lifelong, persistent learning and memory problems).

So, yeah, too much alcohol is bad.

But here's the thing about alcohol: in moderation — no more than two drinks per day for men or one per day for women — it appears to modestly reduce the risk of heart disease, stroke, diabetes, and death according to the Harvard School of Public Health. Psychologist Jennifer Wiley of the University of Illinois even published a study in 2012 showing that a few drinks can actually improve people's ability to find creative solutions to puzzles, especially increasing the chances that a person will have novel insight. And these benefits aren't reserved to one type of alcohol. Red wine has not in fact been proven to have benefits beyond those associated with other sources of alcohol. Nor has resveratrol, a substance found in red wine, been shown in scientific clinical trials to benefit humans.

All in all, we are left with evidence that is modest, at best, to justify the use of moderate alcohol. What's beyond doubt, on the other hand, is that more brains and lives are ruined by overuse of alcohol than by any other substance. For this reason, I wouldn't call it a smart drug.

NEURO BUSTED: 12-STEP PROGRAMS ARE *NOT* THE ONLY POSSIBLE TREATMENT FOR ALCOHOL ABUSE

If you or a loved one is struggling with alcohol use, I implore you to take a look at an article published in the April 2015 issue of the *Atlantic* by journalist Gabrielle Glaser. Intensely reported, "The Irrationality of Alcoholics Anonymous" examined how AA has come to be considered (in the United States, at least) the *only* effective treatment for alcohol use disorder. Yet recent studies show that less than 10 percent of people who enter AA achieve sobriety. By contrast, Europeans favor straightforward medical management that is far more effective, including cognitive-behavioral therapy and the use of naltrexone to block craving or Antabuse to cause nausea when taken with alcohol. Although FDA-approved and demonstrated in randomized clinical trials to significantly reduce drinking, these medicines are used by fewer than one in a hundred Americans with a drinking problem to treat their addiction. NPR's *Radiolab* has aired a compelling segment on the effectiveness of medications compared to that of 12-step programs for treating addictions. It covers the topic well.

CAFFEINE

Found in coffee, tea, soda, and energy drinks, caffeine is the most widely used psychoactive drug in the world. In the United States, 85 percent of adults drink some kind of caffeinated beverage. A stimulant to the central nervous system, it reduces fatigue and drowsiness and improves reaction time, mood, concentration, and physical coordination. So pronounced are the effects that many professional athletes, especially those involved in endurance sports, take caffeine at regular intervals during competitions.

But the effects of caffeine on learning and memory are unclear, with most studies suggesting no direct benefit. Most reviews of the scientific evidence conclude that at lower doses it doesn't help and at higher doses it hurts. Yet because people tend to feel better and more alert after ingesting caffeine, many think of it as a kind of smart drug.

A fascinating study published in 2015 explored this paradox. One gave college-age people either a dose of the widely sold 5-Hour ENERGY shot or a look-alike, taste-alike placebo without caffeine. Incredibly, 90 percent of the participants *thought* they were performing better on tests of short- and long-term cognitive functioning. Yet on average, those who got the caffeinated version performed no better than those who received the placebo.

However, in a different study in 2016, forty-three participants were given caffeinated coffee or decaffeinated coffee and evaluated with tests related to memory and executive planning. After review of the subjects' performances on assigned tasks, the authors concluded that "performance was significantly improved on planning, creative thinking, and memory." For this reason, I would call it a smart drug.

COCAINE

When my oldest son was in ninth grade, I took him with me for a week in La Paz, Bolivia. He brought books and games for the children at Hospital del Niño; I brought special equipment to perform brain surgeries.

At nearly 12,000 feet above sea level, La Paz is the highest capital city in the world. On previous visits, I had suffered from splitting headaches for days due to altitude sickness. This time, however, I decided to handle it the way the locals do. My son and I went to the market in the Villa Fátima neighborhood and bought coca leaves to chew, with a touch of lime.

Yes, coca is legal in Bolivia. According to the country's revised constitution, passed by referendum in 2009: "The State shall protect na-

tive and ancestral coca as cultural patrimony, a renewable natural re-
source of Bolivia's biodiversity, and as a factor of social cohesion; in
its natural state it is not a narcotic."

Chewing coca leaves does not get you high. It's more like coffee,
a mild stimulant that can suppress fatigue, hunger, pain — including
headaches. Plus, in a ninth-grade boy off the grid with his father, I
thought the coca leaves might produce a feeling of getting away with
something cool and memorable, as well as an authentic moment to
have "the Talk" about smart and stupid drugs.

The effect of the natural leaves is in no way comparable to the effects of cocaine, the active ingredient. After first being synthesized by
German chemist Friedrich Gaedcke in 1855, cocaine was soon seen
as a miracle drug, a breakthrough treatment for blocking pain during
surgery. Sigmund Freud even wrote an 1884 treatise, "Über Coca," de-
scribing it as a treatment for morphine addiction. Soon it was being
used as an ingredient in medicines, including one called Coca-Cola.

And then everyone, including Freud, began to realize that cocaine
was terribly addictive, that it could cause paranoia, agitation, high
blood pressure, and rupture of arteries in your brain. In 1914, it was
banned except by prescription.

To this day, neurosurgeons like me still use it on rare occasions —
not on ourselves, of course, but on our patients, to narrow the blood
vessels in the lining of their nostrils when we use the nose to access
the base of the brain.

Otherwise, coke is a nightmare, even if some continue to flirt with
it for a momentary kick. It can damage blood vessels and raise blood
pressure, potentially causing headaches, convulsions, heart attacks,
and stroke. Not a smart drug.

DIETARY SUPPLEMENTS

People who call themselves "brain hackers" like to take all kinds of
so-called natural substances that they think will make them smarter.

However, there is no convincing scientific evidence that aniracetam, ashwagandha, *Bacopa monnieri,* carnitine, huperzine A, omega-3, or any of the products sold as dietary supplements actually improve cognitive functioning. If the manufacturers could prove it in randomized clinical trials, they would provide those studies to the FDA to get them approved as prescription medicines and reap billions of dollars in profits. Instead, they slip under the special rules passed by the U.S. Congress for dietary supplements, which require no proof of either safety or efficacy, and are satisfied making millions.

The one thing that has been proved about dietary supplements, over and over again in studies by *Consumer Reports, Journal of the American Medical Association,* and others, is that they often have substances not listed on the ingredients label and higher or lower doses of what *is* listed. Many of them also have dangerous or adverse side effects, such as an increased risk of heart attacks, high blood pressure, weight gain or loss, dry mouth, frequent bowel movements, belching, nosebleeds, and a fishy taste in the mouth.

Take, for example, *Bacopa monnieri,* a medicinal herb used for thousands of years in Ayurveda, the traditional medical system of India, as a treatment for epilepsy, asthma, ulcers, tumors, leprosy, anemia, enlarged spleen, gastroenteritis, and more. A study in male mice found that it reduced their fertility and sperm count. And while some continue to claim it can improve memory, no good studies have ever demonstrated that.

This book is not the place to offer a thorough examination of the bizarre world of dietary supplements, but before moving on, I want to share a story about one supplement that is no longer available because it was found to be too dangerous. Ephedra, an herb used for centuries in traditional Chinese medicine, gained popularity in the 1990s as a mild stimulant to improve focus and aid in weight loss. When I was in college, many people I knew used it to help them study. I used it a few times, too, particularly during lengthy exams. It definitely felt like it improved how long I could focus. In those days, you could buy it at gas stations, convenience stories — anywhere. Then some professional athletes dropped dead while using it as did others because it raised blood pressure, quickened heart rate, increased sensitivity to heat stroke, and did pretty much all the other things that any amphetamine-like substance will do. Adverse effects reported to the National Poison Data System peaked at 10,326 in 2002, with cases significant enough to require hospitalization hitting 108 that same year. In 2004, the FDA banned it.

MARIJUANA

As of this writing, nine U.S. states have made recreational marijuana use legal, while more than half have legalized it for medical treatments only. As medicine, marijuana or its constituent ingredients have been found to be effective — for some people — in reducing pain, nausea, vomiting, epileptic seizures, post-traumatic stress disorder, and perhaps even sleep apnea.

Back in the 1960s, marijuana was promoted as mind-expanding, a claim seemingly supported by its use among free-thinking hippies, professors, actors, musicians, writers, Beat poets, plus that couple who lived down the block and was always throwing cool parties.

But cannabis has some worrisome side effects that its advocates, then and now, prefer to sweep under the rug. Like, for instance, car crashes. A study of U.S. traffic accidents found that the rate of crashes rose by 12 percent on April 20 (420 is the marijuana culture reference), the date widely celebrated as "Weed Day."

That shouldn't be surprising, because countless studies have found that while high on pot, people have reduced attention and reduced performance on psychomotor tasks. Short-term effects on memory and decision-making are also well established and fairly obvious to regular users (or to anyone who ever listened to a recording by the comedy duo Cheech and Chong).

Some studies have also found that chronic use of cannabis during adolescence results in a lower IQ and long-term thinking deficits. But others have concluded that it's actually the other way around: people with lower IQs are the ones more likely to become stoners.

Of greatest concern is the growing body of evidence linking regular marijuana use to an increased risk of developing severe psychiatric illnesses, especially during adolescence. In 2017, just over 37 percent of twelfth graders used it at least once during the year, and 5.9 percent used it every day — a huge jump over 1992, when only 1.9 percent were daily users. The more regularly a teen uses marijuana and the higher the potency, the greater his or her risk of becoming schizophrenic. Heavy users are also more likely than others to be depressed; and, what's worse, marijuana use *during* depression reduces the rate of recovery.

But let's not exaggerate the risks. It is one of the rare drugs that has had no cases of overdose-related deaths. Most people who occasionally use marijuana do just fine.

Marijuana might make you feel more relaxed, make you laugh, enjoy food and music more, and it certainly has some proven medical benefits. Whether it makes you creatively insightful is hard to pin down (especially the morning after). But calling it a smart drug is a bit of a stretch.

MODAFINIL

Approved by the FDA in 1998 as a treatment for excessive daytime sleepiness due to narcolepsy, shift work, or sleep apnea, modafinil is generally considered one of the most effective and least dangerous "smart" pills out there. Originally sold under the brand name Provigil, it is also now available as a less costly generic drug, but a doctor's prescription is still necessary.

Dozens of studies have been published on its cognitive effects. A 2012 study of thirty-nine sleep-deprived doctors, for instance, found that those given modafinil rather than a placebo performed higher on tests of memory and planning after staying up all night and were less impulsive in making decisions.

Even in people who have had a normal night's sleep, studies have found that a dose of modafinil not only helps on tests of planning, memory, and creativity but also enhances enjoyment on cognitive tasks.

Still, not all studies reached positive conclusions; so in 2015, researchers at Oxford University reviewed twenty-four papers investigating modafinil's effects on healthy, non-sleep-deprived adults. They concluded that studies testing simple mental tasks had reached mixed results, but that those testing more *complex* cognitive abilities found that modafinil consistently enhanced attention, executive functions, and learning with few side effects.

When I taught biology to undergrads at the University of Southern California, I asked the class how many were taking modafinil. Easily a third said they had tried it.

I will neither condemn nor endorse such use. The one thing I will say is that, at least, modafinil does not appear to kill or cause the kinds of serious side effects that ephedra did before it was banned.

NICOTINE

Cigarette smoking causes nearly half a million deaths per year in the United States. That's about one in five deaths overall. So tobacco is toxic. I work at a cancer center. Enough said.

And yet nicotine *without* tobacco is a really, really interesting drug with all sorts of fascinating effects on the brain.

In 1966, a huge study conducted by Harold Kahn at the National Institutes of Health found all the expected harms of smoking—greatly increased rates of cancer, stroke, heart disease, lung disease, and death—but also something very odd: smokers were three times *less* likely to develop Parkinson's disease than nonsmokers. Follow-up studies have since confirmed beyond doubt that tobacco users have a significantly decreased risk of developing the movement disorder.

Why? Nicotine, like many other addictive drugs, increases the brain's levels of dopamine — a neurotransmitter that can boost attention and reward-seeking behaviors but also facilitates smooth, voluntary movements.

More recent studies by neuroscientist Maryka Quik, program director of the Neurodegenerative Diseases Program at the research center SRI International, have shown that animals with movement disorders improve significantly when given nicotine. Two large studies of people with Parkinson's are testing whether that effect holds up in humans.

Other studies have found that nicotine improves attention and memory and that it might slow or delay the progression of mild cognitive impairments, which often precede a diagnosis of Alzheimer's disease. One study even found that a low-dose nicotine patch reduces impulsivity and improved memory in people with ADHD.

Could nicotine turn out to be a smart drug? More importantly, since nicotine patches and gums are available without a prescription, should you or a loved one consider trying a low dose? Here is what Paul Newhouse, director of the Center for Cognitive Medicine at Vanderbilt University School of Medicine in Nashville, told *Neurology Today* in 2012: "It looks safe. It looks reasonable. But the long-term clinical impact still has to be better assessed. I think it's still a little premature. But certainly the published data suggest it has potential benefits for [mild cognitive impairment] and it definitely appears to be a reasonable strategy for further testing. I think the risks are fairly low. But I'd be very clear with a patient to say there is not enough information to say this has been proven effective."

To see for myself what the nicotine patch would be like, I gave it a try for a few weekends when I was working on this book. I even tried to compare it to a placebo by asking my wife to put either the medium-strength patch or an ordinary Band-Aid on my back shoulder, where I couldn't see it.

On the days when she applied the real thing, I could definitely feel it. It kicked in like a cup of coffee: The cobwebs went away, and the writing came more easily. But it didn't fade like coffee after a couple of hours, and I never felt jittery. The patches are meant to last for twenty-four hours, and certainly they gave me four hours of writing that felt effortless.

Would I recommend it? Hard to say. If you ask your doctor about it, he or she will likely say you're crazy to even consider it. The patches are surprisingly expensive, but they are available without a prescription. Beyond that, if you're a healthy adult, it's something to consider.

PRESCRIPTION STIMULANTS

Recent national surveys of U.S. students have found that more than 6 percent of all high school seniors used a prescription stimulant at least once during the year to help them study, compared to 10 percent of college students overall, and 20 percent of Ivy League students. Not smart. When used illicitly as a study aid, prescription stimulants are guaranteed to cause serious adverse events, or addiction, in a small but significant portion of the millions of kids and adults who try them.

The two major stimulants prescribed to children and adults with ADHD are Adderall (benzedrine) and Ritalin (methylphenidate). Both have been prescribed for decades as a usually safe and easy way to help people study who otherwise would be unable to settle down and concentrate. Certainly, they are overused as a disciplinary tool. Some children find the side effects unacceptable (including headaches, stomachaches, loss of appetite, and difficulty sleeping), and some parents adamantly oppose their use in any situation. But for others, when prescribed and monitored by a diligent physician, these medicines can make a huge difference.

For people without ADHD, both Adderall and Ritalin will likewise improve concentration and attention. Need to cram for an exam? One dose and you're good to go for the night.

Fantastic, right? Sure, unless you're concerned about becoming paranoid, a common side effect from the excessive use of stimulants like these. The British military stopped allowing its soldiers to use stimulants more than fifty years ago because they realized that paranoia in soldiers was not a good thing.

And no study has ever shown these drugs to be true "cognitive enhancers." They do not make you smarter; they just keep you awake and alert longer. Although, as with coffee, studies have found they make people *feel* smarter. For instance, a placebo-controlled, double-blind study published in the journal *Neuropharmacology* looked at the effects of Adderall on thirteen measures of cognitive ability in healthy people. "The results did not reveal enhancement of any cognitive abilities," the study found. But, it added, "participants nevertheless believed their performance was more enhanced by the active capsule than by placebo."

Unless a doctor has diagnosed you as having ADHD, try getting some sleep instead of messing with your mind with prescription stimulants.

NEURO GEEK: ALZHEIMER'S DRUGS FOR THE HEALTHY?

While most adults will *not* develop Alzheimer's, many aging adults will have mild cognitive impairments that affect how they perform in daily life and work. In the sector of the scientific community focused on developing drugs to improve the symptoms associated with dementia, there is an understanding that those same drugs may be used by healthy people with the goal of maintaining cognitive acuity or even, in certain circumstances, improving mental agility above their established baseline.

One study from Stanford University examined just that question: Could otherwise healthy, middle-aged individuals improve their cognitive performance using agents designed to combat dementia? They invited airplane pilots with an average age of fifty-two who were interested in participating and evaluated performance in flight simulations. Every three minutes, the pilots were peppered with complex commands and codes that had to be remembered and implemented while under stressful flying conditions. After seven simulations, the study subjects were given Aricept for a month and then retested under equally demanding but new simulations. The results showed that they performed better in the later round of evaluations, and the authors stated that Aricept "appears to have beneficial effects on retention of training on complex aviation tasks in nondemented older adults," and these findings were ultimately published in *Neurology*.

There is already a major pharmacologic market for physical performance-enhancing drugs. With an aging population, interest in medications for cognitive enhancement will likely grow.

6

SLEEP ON IT

Part of my workday as a neurosurgeon is spent in the neurosurgical intensive care unit, or neuro ICU, where I watch over patients recovering from surgeries and battling to survive life-threatening brain injuries. Most of them are kept sedated and on ventilators — some, so deeply sedated they are just this side of death. But don't mistake the state they're in for sleep.

Sleep is a firestorm of brain activity. Instead of taking in new information, our brain's subconscious is occupied defragging, deleting, and storing the prior day's doings for long-term retrieval; cleaning out bits and pieces of discarded brain schmutz; and presenting us with immersive 3-D virtual stories in which we are the star.

Rest, you call that? The brain never rests. So essential to life are the myriad activities the brain engages in during sleep that without it, we die.

And there's the rub. The brain's refusal to settle down, awake or asleep, is a problem for my sickest patients.

Such was the case with a patient named Emily. I first met her in the fall of 2004, after returning from a week off following the birth of my second son. The evening before my first day back, I dropped by the neuro ICU to "run the list"; that is, to review the status and needs of each patient. That night, the long, L-shaped neuro ICU was near capacity, with nearly all the patients on ventilators. Emily, eighteen years old and a freshman in college, was the one who demanded most of my attention.

Three weeks before, the car in which she'd been a passenger was in a head-on car collision. Although she'd been wearing a seatbelt and the airbag deployed, the whiplashing force of the extreme deceleration had snapped an unknown number of her brain's axons and dendrites while swelling and popping neurons like so many water balloons. Those microscopic tears and explosions, in turn, caused her whole brain to swell and throb, just as your nose would swell if punched.

A small catheter had been placed into her brain to measure her intracranial pressure. Normal pressure for a healthy person is between 7 and 15 mmHg. Anything over 20 mmHg is considered dangerous and causes an alarm to go off in the neuro ICU, as it can result in the crushing of brain cells, the smashing of whole sections of the brain against their cranial confines, and, eventually, death.

Emily was already sedated with propofol and a morphine drip, not to mention a paralytic agent to keep her from moving. She also had two pancake-sized pieces of her skull removed, one on each side, to give her brain space to expand. But when I arrived that evening, her pressure was creeping toward 21.

Her young age was both good and bad. Good, because the young demonstrate the most surprising and unexpected recovery. Bad, because they have naturally engorged brains that are most prone to fierce swelling.

Knowing that her life and future hung by a thread, I decided to try the only, but dangerous, option in cases such as this. I walked close to the nurse and softly said, "Barb her."

Pentobarbital would suppress the electrical spikes in her brain by putting her into a coma-like state. Some occasional slow brain waves would still be seen on her EEG, but not the faster, sharper spikes and bursts that are normal whether a person is asleep or awake. Eliminating those spikes would bring her brain to its lowest metabolic demand and make it biologically dormant while the swelling abated over the next few weeks. Essentially, I was flatlining her, putting her in a station near death, in order to protect her from actual death.

The monitors beside her bed showed a grid of data, including heart rate, breathing rate, and heart rhythm, but all I cared about now were her intracranial pressure and the waves on the EEG.

Sure enough, over the course of hours, the spikes stopped and her pressure dropped. Now instead of being just over 20, her intracranial pressure was just below. Not ideal but not deadly. She wasn't sleeping and wasn't dreaming, but her brain was finally getting the electrical and metabolic quiet it needed.

For weeks, we kept her in a bed that works like a giant rotisserie, slowly rotating her so that her lungs wouldn't fill with fluid and she wouldn't develop bed sores. Her pressure held below 20, and scans of her brain never showed the telltale dark swatches of neural death.

Finally, one by one, we began tapering off the doses of her medicines. First the pentobarbital, which took weeks to clear from her system, then the paralytics, then the sedative, then the narcotic (but not completely because I didn't want her to be in pain as she regained consciousness).

Nearly two months after I had first seen her, Emily began moving her arms, hands, and legs. As she slowly came to, her fidgeting was reassuring. And finally, one morning she opened her eyes ever so slightly. Her parents' shouts brought a nurse rushing over, and soon

I was made aware. I asked Emily to stick out her tongue. She did. I asked her to give me a thumbs up. She did. She was awake and could understand and respond.

A year later, her parents sent me a letter. Emily was doing okay and was working at a coffee shop, they wrote, "despite being asleep for all those months."

Interesting, I thought: She *wasn't* asleep at all for those months. Sleep is not a suppression of brain activity. Quite the opposite. Sleep calls on deep powers of the brain never used during wakefulness.

It might seem odd for me to have opened this chapter on sleep with a story about someone who *wasn't* sleeping, but the popular notion that sleep is a time for the brain to rest is just wrong. The only people whose brains are truly resting are those who are placed in a chemical coma, like Emily. They do not dream, and they are not asleep as doctors understand sleep.

So, let's look now at what really goes on when you're slumbering.

WHY WE SLEEP

What *is* sleep, anyway, and why do we do it? Since the dawn of the twenty-first century, neuroscientists have learned that not only do all mammals do it but so do all fish, birds, bees, worms, flies, and ants. (Queen ants sleep about nine hours per night, one study found, while worker ants get half that much but supplement it with a bunch of power naps.) Even jellyfish without a brain have been observed sleeping — or at least doing a pretty good imitation of it.

The great mystery is *why* most creatures on Earth allocate part of their life to sitting still: not eating, not drinking, not chasing a mate, and not keeping up their guard against predators. What is so valuable about sleep that it's worth giving up all the benefits of wakefulness?

One thing we know that humans and other mammals do during sleep is to transform short-term memories stacked up during the day into memories that can last a lifetime. Although the hippocampus is essential for making new memories, other brain regions distributed across your cortex are necessary for long-term storage. So, you can think of sleep as the time when your brain ships memories from your hippocampal loading dock to distant corners across your brain.

After studying for a test, students will actually remember more after a nap or a night's sleep than if they had stayed awake and kept studying for an extra few hours. A study in the journal *Nature* found that sleep can even increase problem-solving insight. After allowing participants to grapple briefly with a puzzle, German researchers found that those who then slept before tackling it again were three times more likely to solve it than were those who simply tried it later in the day without sleeping.

Likewise, people who try learning a dance move or any other physical skill perform better the next morning than they did immediately following the training.

Of course, not all short-term memories and insights go into long-term storage. Quite the opposite. Most of what happened yesterday, let alone last week or last year, is literally deleted from your brain at night. As a recent, brilliant paper in the *Journal of Neuroscience* put it: "One of the essential functions of sleep is to take out the garbage, as it were, erasing and 'forgetting' information built up throughout the day that would clutter the synaptic network that defines us." It's not that your unnecessary memories simply fade like old photographs that have been left out in the sun too long, but that they are actively erased during sleep. What's more, the paper stated, "This targeted forgetting is necessary for efficient learning, and deficits in this process may underlie various kinds of intellectual disabilities and mental health problems." And taking out the garbage is not just a metaphor for eliminating neural connections. A growing body of research shows that during sleep, the brain literally flushes out cellular detritus built up during the day.

RAPID EYE MOVEMENT

No doubt you have heard about REM (rapid eye movement) sleep, when the eyeballs can be seen darting to and fro beneath a sleeper's lids. REM was discovered in December 1951 by a graduate student named Eugene Aserinsky when he asked his eight-year-old son to spend a night in the sleep laboratory where he was working as a researcher. During multiple occasions that night, the boy's brain waves began going haywire, displaying not only the slow delta waves characteristic of deep sleep, but the more rapid waves of beta, alpha, and theta associated with wakefulness. At first Aserinsky thought the boy had awakened, but each time he went in to check on him during the sudden burst of brain activity, the boy was fast asleep. What's more, his eyes were dancing around beneath the lids.

It took two more years for Aserinsky and his advisor to realize that these periods of REM were associated with dreaming, because when they awakened people during REM, they almost always reported being in the middle of a vivid dream. Ever since, a myth has arisen that dreaming happens *only* during REM.

In fact, dreams occur through much of the night even without REM. Some studies have found that non-REM dreams seem to be less vivid and emotional than those experienced during REM, but other studies have found no such difference. When French researchers suppressed REM sleep with a drug, they found that "long, complex and bizarre dreams persist." In another study, researchers from Finland and Wisconsin awakened participants repeatedly throughout the night and found that when they were in non-REM sleep, they had nevertheless been dreaming more than half the time.

THE MISINTERPRETATION OF DREAMS

So, what is the purpose of dreams? Once upon a time, dreams were taken as omens from above. And then along came Sigmund Freud, who asserted, beginning with his 1899 book *The Interpretation of*

Dreams, that they are the symbolic manifestation of our fears, desires, anxieties, and repressed childhood memories. Dreams, in Freud's view, are a kind of Roman Colosseum where sexual desires and other wishes for supremacy (the part of the personality he called the "Id") do battle with that part of our personality that tries to keep a lid on the Id — the seat of censorship and conscience that he called the Super-Ego.

Brilliant and insightful, Freud's theories and observations catapulted psychology into the twentieth century, much as Einstein's theories did for physics. But where Einstein's theories remain a bedrock of modern physics, his formulas and predictions demonstrating uncanny accuracy, Freud's ideas have held up less well. Nowhere in the brain can you find an Id, an Ego, or a Super-Ego. And his claims about the meaning of dreams have found no scientific support.

This is not to say that dreams aren't cool and fascinating. Sometimes they do reveal an unexpected solution to a vexing problem or at least a fresh insight. And every once in a while, they provide artists and scientists alike with a poem, a formula, a song — some new creation. Scientifically speaking, though, we still do not really understand *why* we dream — not in the way we understand why we breathe, eat, and have sex. It remains, as the journal *Science* noted in its 125th anniversary issue, one of nature's great unsolved mysteries.

HOW MUCH IS ENOUGH?

In 1995, sleep researcher Allan Rechtschaffens devised an experiment to see what would happen if he kept rats awake for a prolonged period. After a few days, their body temperatures began to fall, and they started losing weight even though they were eating more food. Ulcers began appearing in their tails and on their paws. After a few weeks, they all died.

In people, an exceedingly rare and incurable genetic disease called fatal familial insomnia causes a total inability to sleep when an affected individual reaches a particular age, often their fifties. After six to thirty months of no sleep, they die.

So while we might not understand *why* we need sleep, it's clear that we do. But how much? As a surgical resident in the 1990s, I and others were required to work shifts that lasted up to 40 hours a pop, with workweeks for surgeons in training routinely running 120 hours. But studies showed that sleep-deprived doctors made significantly more medical errors than did those with more rest. As a result, beginning in 2003, the organization overseeing medical education in the United States set strict limits on how long residents could go without sleep. The limits were set at 80 hours per week for medical and surgical trainees, with an exception of 88 hours allowed for neurosurgery residents. (They initially tried to cap neurosurgery residents at 80 hours, like everyone else, but because there are so few of us, hospitals were left short-staffed when emergency brain trauma cases showed up. I suppose they realized that the only thing worse than a tired young neurosurgeon when someone's brain is bleeding is no neurosurgeon at all.)

Still, some people insist they get along just fine on as little as four or five hours of sleep per night — in fact, some business executives and college students boast about how little sleep they get away with. "I'll sleep when I die," they say.

More likely, they will die from lack of sleep. A 2010 analysis of sixteen prior studies involving over 1.3 million people found that those who averaged less than six hours of sleep per night were 12 percent more likely to die before age sixty-five than those who slept six to eight hours a night. However, the same study found that people who slept more than nine hours of sleep per night had a 30 percent increased risk of early death.

Other studies have found all kinds of health risks associated with sleeping too much or too little. The Nurses' Health Study, for instance, found that women who slept more than nine hours per night had a 38 percent increased risk of heart disease over those who slept eight hours. And a study from the American Heart Association found that people who already have metabolic syndrome (increased weight, blood sugar, and blood pressure) will *double* their risk of death if they sleep less than six hours per night.

Of course, the need for sleep varies dramatically over the lifespan, with most babies and toddlers devoting half of each twenty-four-hour period to sleep, while adults over the age of sixty-five do well with as little as seven hours of sleep per night. I, like many, struggle to sleep well.

Take a close look at the latest recommendations from the National Sleep Foundation.

SLEEP DURATION RECOMMENDATIONS

AGE	RECOMMENDED	MAY BE APPROPRIATE (LOW AND HIGH)	NOT RECOMMENDED
School-aged children 6–13 years	9–11 hours	7–8 hours 12 hours	Less than 7 hours More than 12 hours
Teenagers 14–17 years	8–10 hours	7 hours 11 hours	Less than 7 hours More than 11 hours
Young Adults 18–25 years	7–9 hours	6 hours 10–11 hours	Less than 6 hours More than 11 hours
Adults 26–64 years	7–9 hours	6 hours 10 hours	Less than 6 hours More than 10 hours
Older Adults ≥ 65 years	7–8 hours	5–6 hours 9 hours	Less than 5 hours More than 9 hours

In particular, pay close attention to the levels that are *not* recommended. While the recommended amounts are ideal, it's the "not recommended" levels that are cause for concern.

NEURO BUSTED: YOU CAN NEVER CATCH UP ON SLEEP

When laws were passed to limit how long medical residents could work without sleep, most of them included an additional requirement that residents get at least four full days off every month. My program director at the time offered us the option of taking off one day every week. But the residents turned him down on that one. I knew it was the *second* day off — two in a row — that led to restfulness.

The first morning off after a 120-hour week, there are too many cobwebs. Only after the second morning of not setting an alarm clock did I feel the clarity and calm that sleep can provide: that feeling of being human again. Only then did I feel I had caught up on sleep. So in order to get that restful feeling that came with the second morning of sleeping in, I elected to work twelve days straight to get a two-day period off.

Despite a long-standing and pervasive notion that one cannot catch up on sleep, new research shows that you *can*, which fits with my experience. And while the bodily stress endured from nearly a decade of sleep deprivation has likely taken its toll on my health, this study is reassuring in that at least some of the health risks accrued were partially mitigated by those two-day off periods. Ultimately, weekends do work as an effective way to catch up on sleep, so protect *both* weekend mornings if you're sleep-deprived during the week. I do.

DARK AT NIGHT, LIGHT BY DAY

Allow me to shed some light, so to speak, on a factor that can affect not only the length and quality of your sleep but your overall health: too much light at night and not enough during the day. We have a built-in timing device in our hypothalamus. (I called it a surgical no-fly zone in chapter 1 because of its importance.) It sits in the center of your brain and has a tiny cluster of about 20,000 very specialized neurons that receive direct input from your eyes. These neurons (collectively referred to as the suprachiasmatic nucleus) tell the hypothalamus about changing night-day cycles. The hypothalamus then processes this information to guide how it regulates behavior, hormone levels, sleep, and metabolism.

Three scientists discovered some of the genes that mediate our biological rhythms on a twenty-four-hour cycle and were awarded the Nobel Prize in 2017. The Nobel committee said that they were "able to peek inside our biological clock," explaining how plants, animals, and humans adapt their biological rhythm so that they are synchronized with the Earth's revolutions.

Disruption of this biological rhythm is closely linked to a wide range of diseases including obesity, type 2 diabetes, depression, and even cancer. David E. Blask of the Laboratory of Chrono-Neuroendocrine Oncology summarizes it well. "We evolved to see bright blue, full-spectrum light during the day and to have complete blackness at night," Blask says. "Both are really healthy for your circadian system. It's all about this oscillatory balance under natural conditions between light and dark."

That said, none of us is going back to living in caves without electricity, and relatively few jobs these days involve outdoor work. But to get a better night's sleep and improve your health overall, I recommend that you avoid bright light at night to the degree possible and get outside for at least twenty minutes per day to enjoy the sunshine. If you can't get outside for whatever reason, get yourself some bright, white, full-spectrum lightbulbs. There's a reason, after all, why the word *gloom* is used to describe both a lack of light and a state of mind!

INSOMNIA

The term *insomnia* is thrown around loosely, but according to the latest definition from the American Psychiatric Association, five criteria must be met for a diagnosis:

1. You are dissatisfied with the quality or quantity of sleep you are getting, due either to difficulty initiating or maintaining sleep, or waking up early and being unable to fall back asleep. So if you're not dissatisfied — even if you fit these other criteria — you don't have insomnia.
2. The lack of sleep is causing significant distress or impairment in your work or personal life, whether in your behavior or your emotions. So even if you are dissatisfied with the amount of sleep you're getting, if it's not causing actual problems, it's not insomnia.
3. Your difficulty with sleep has lasted at least three months and occurs at least three times a week.
4. Your difficulty continues even though you have ample opportunity to sleep. So it's not caused by work or other demands.
5. The lack of sleep is not better explained by another physical or mental disorder.

Estimates of how many people have insomnia have been all over the map, with some loosey-goosey studies claiming that up to one-third of all adults have it, which is just ridiculous. The best study I know of was published in 2014, involving over 40,000 respondents to a major health survey in Norway. (They even went back to the people who did *not* respond to the survey and got another 7,000 of them to answer two questions about sleep.) They concluded that 9.4 percent of women and 6.4 percent of men qualified as having insomnia under the strict criteria listed above. But people who reported that their overall health was "very bad" were eight *times*–that is, 800 percent — more likely to suffer from insomnia than were those whose health was "very good."

MY EXPERIENCE WITH SLEEP DEPRIVATION

Each one of my patients undergoes episodic sleep deprivation after brain surgery. This fragmented sleep is the consequence of a routine clinical practice in ICUs across America. We ask the nursing staff to examine post-operative patients every hour on the hour to ensure they are able to rouse, talk, and move. The formal nursing order is: *Q1hr neuro-checks.*

While it gives us vital information about the brain's connectivity, the resulting disrupted sleep pattern is a major contributor to pushing patients' minds into a delirium termed ICU psychosis. Although these patients usually tally six to eight hours of total sleep, it's not continuous, so delirium emerges, and patients may experience disorientation, agitation, and halucinations. After the initial stay in the ICU, patients are transitioned to step-down areas of the hospital where vitals and neuro-checks are reduced to every four hours. And most will have a complete resolution of their delirium as medicines are removed and more prolonged sleep is permitted.

The hazards of fragmented sleep are part of surgical lore as well. When asked about our thoughts on sleep, as residents we had a phrase that encapsulated it well: "Patients don't sleep well and we don't sleep at all." Whereas the disrupted sleep of patients was something we unintentionally prescribed by monitoring them for post-operative brain hemorrhages, our own sleep deprivation was generally by choice. Why?

Usually, the demands of the hospital led to nights when the incessant workload simply allowed no rest. But on some nights, there were windows when one could sneak in a bit of sleep. It didn't take more than the first few nights of being "on call," however, to appreciate why it was often preferred by the more experienced among us to stay awake all night instead of giving in to the seduction of an hour or two of sleep. Simply, just as the ICU patients experienced, for the surgical residents, fragmented sleep often felt worse than no sleep at all. And the morning after was still a full fourteen-hour work day that required as much focus as one could muster.

So, in my case, I would force myself to endure self-imposed insomnia by spending the nights patrolling the sickest patients in the ICU and hanging out with the technicians and nurses that were on their graveyard shifts. Even now, as those forty-hour shifts are behind me, I would rather sleep well for five hours straight than get seven hours of disrupted sleep in short bursts with interruptions.

LUCID DREAMING

Have you ever had the experience of waking up inside a dream so that you're still sleeping when you suddenly realize that you are actually dreaming? Lucid dreaming, as it's called, is outrageously cool and fun. Sometimes it occurs during that transitional stage between sleep and wakefulness. But it can also happen during ordinary sleep. With practice, you can learn how to have such dreams and how to control them.

I learned about lucid dreaming from an outstanding book that was first published in 1975: *Creative Dreaming* by Patricia Garfield, PhD. Its instructions for inducing lucid dreams work as well today as they did then. She and other researchers have found that lucid dreams can even help people who suffer from chronic nightmares by permitting them to deal with the frightening scenarios and menacing figures in a creative, thoughtful way.

Here's a two-week plan for lucid dreaming. It's best done during a vacation because sleeping in late is an important component:

1. Every day and night, tell yourself repeatedly that you will have a dream in which you fly like a bird. Because it's so obviously impossible, flying is one of the most common reasons that people

realize they are dreaming. By repeating in your mind "Tonight I will fly" as often as possible, you increase the chance that you will have a flying dream. After all, we all tend to dream about whatever most occupies our mind, whether work, family — or flying.

2. Allow yourself to sleep as late as possible each morning. The longer you sleep, the more vivid dreams tend to become, and the better the chances that you will remember them.

3. When you awaken, do not roll over or get up. Keeping your head still, lie quietly and try to remember your dreams. The longer you try, the more you will remember. Spend at least ten minutes on this every morning.

4. Keep a dream log. Once you have remembered as many of your dreams as possible, write them down in a journal.

If you don't end up having a few lucid dreams, at least you will enjoy some good, long nights of sleep and the entertainment of your brain's awe-inspiring capacity for interactive virtual reality.

NEURO GYM: HOW TO GET MORE ZZZZZ'S

What do you do if you or a loved one has insomnia? Some people buy a bottle of cold medicine and take a couple shots. But over-the-counter drugs containing an antihistamine work only to *initiate* sleep; they don't help you stay asleep through the night. More importantly, they are *not* recommended — and *do not work* — as a treatment for chronic insomnia.

Neither does whiskey, beer, gin, vodka, or wine. And anyway, drinking yourself to sleep causes more problems than it solves. At most, alcohol is a sleep initiator. A night cap is not going to help with deep, restful sleep.

If you prefer a natural treatment, the hormone melatonin is sold without prescription, but the power of melatonin on sleep induction has been greatly exaggerated. Our brain's pineal gland is the source of melatonin, and there is extensive experience with the surgical removal of this anatomical structure. In a prospective study, patients were evaluated before and after pineal gland removal, and the results were clear: "After pinealectomy, melatonin was markedly diminished, mostly below detection limit. Sleep-wake rhythm during normal daily life did *not* change."

Even prescription drugs like Lunesta, Ambien, and Sonata were lately found by *Consumer Reports* to help people fall asleep a mere eight to twenty minutes faster. Even then, there is a significant risk of side effects. People occasionally report feeling sleepy or having a headache the day after. Most infamously, some people taking Ambien have gotten out of bed, stepped into their car, and gone driving down the street, all while asleep.

So what do the sleep experts recommend? A detailed list of do's and don'ts is found on the American Academy of Sleep Medicine website, https://aasm.org. I struggle with insomnia myself, so from their exhaustive list, here are some of their recommendations I find most effective and why.

1. KEEP A CONSISTENT SLEEP SCHEDULE. Get up at the same time every day, even on weekends or during vacations. This will help you stay in that circadian rhythm that sets an internal clock for falling asleep. I follow this recommendation only partially, because I'd rather stay up late — and sleep in late — on weekends. But I try to log the same total hours of sleep on the weekend as during the week, or often more.

2. AVOID CAFFEINE IN THE AFTERNOON OR EVENING. Caffeine can stay in your system for ten to twelve hours, so I would say early afternoon is a better cutoff if you are struggling with insomnia. On nonsurgery days, when I'll occasionally have coffee, it's never after midday.

3. IF YOU DON'T FALL ASLEEP AFTER TWENTY MINUTES, GET OUT OF BED. This is a reasonable recommendation. You definitely don't want to toss and turn indefinitely. So I agree that if it's getting close to a half hour and you aren't drifting off, you should consider getting up. But if you do get up, do something mellow and keep the lights dim.

4. USE YOUR BED ONLY FOR SLEEP AND SEX. Well, *only* is a strong word, but the general principle here is good. When I have run out of movies or shows to watch on my laptop, I do like to read in bed at the end of the day. But if I'm struggling with insomnia, I even skip reading in bed because of the light exposure, even on an e-reader. Then my go-to move is to listen to a podcast with the lights off, and, of course, it's got to be something my wife can tolerate.

5. LIMIT EXPOSURE TO BRIGHT LIGHT IN THE EVENINGS. There is no question that nighttime exposure to light disrupts the induction of sleep. Most of the rooms in my own home have a dimmer, and I start turning the lights down around 8 p.m. every evening.

6. TURN OFF ELECTRONIC DEVICES AT LEAST THIRTY MINUTES BEFORE BEDTIME. The hardest recommendation of all to follow. For me, I have to admit, my phone is the last thing I look at each night and the first thing I pick up each morning. I do give myself a "digital sunset" by placing the phone on "night mode" after 8 p.m. That way my teenager, who is increasingly out and about at night, can always get to me. And my neurosurgical colleagues and I always keep our phones on for emergencies that may require some backup.

But if you have tried the recommendations above for chronic insomnia and are still struggling, consider seeking help from a *licensed* sleep therapist who uses cognitive behavioral therapy (CBT) as a first-choice treatment. Using CBT, the licensed sleep therapist will help you learn about habits or attitudes that stand in your way of getting a good night's rest. The therapist will likely suggest that you keep a sleep diary. Some studies suggest that up to 80 percent of people with chronic insomnia get long-lasting help from CBT.

If therapy isn't your thing or you're concerned about the cost, check out an online version of CBT for insomnia at http://restore.cbt program.com/restore or https://www.sleepio.com/cbt-for-insomnia. A study published in *JAMA Psychiatry* found that twice as many people with insomnia were sleeping normally after a year when they used a similar program for six weeks, compared to people who had received routine advice and education: 57 percent versus 27 percent.

NEURO GEEK: ONDINE'S CURSE

Nearly a decade ago, a patient came to me after having been operated on several times by neurosurgeons across the country for a noncancerous cyst that kept returning and pushing on her brain stem. I needed to peel, slice, and dissect the cyst wall off blood vessels so thin they looked like hairs. The surgery went well, and she seemed fine upon awakening. But that evening, as she drifted off to sleep, she stopped breathing. An alarm went off, and the breathing tube that had been in place during surgery was re-inserted. We figured her post-operative narcotics might be suppressing her urge to breathe, a side effect that sometimes kills people taking opiates.

The next morning, she was breathing normally again. We removed the uncomfortable breathing tube, and all seemed well. But that afternoon, when she dozed off, she stopped breathing again. Now it was looking serious.

"I think she has central hypoventilation syndrome," a senior neurosurgeon told me.

Until then, this complication was something I had only read about. Most often caused by an injury to the brain stem, it results in failure to breathe only during sleep and so typically requires lifelong mechanical ventilation whenever the person goes to bed. Interestingly, the disease is also known as Ondine's curse, after a European folktale that has been the source of many literary works, including *The Little Mermaid*. In the original tale (far darker than the animated Disney movie), a water sprite named Ondine is deserted by her mortal husband for another woman. She has her father put a curse on him so that he must now do consciously everything his body once did automatically, on its own. When they kiss goodbye, he forgets to breathe — and dies.

We waited a week to see if the problem would improve as she recovered from the surgery. When it was clear that her condition was permanent, she agreed to undergo a second, simple surgery to get around the need for having a breathing tube placed down her throat every night: a tracheotomy. I made a tiny horizontal incision in the middle of her neck and cut a small opening in her trachea. A hollow plastic hub was positioned into the orifice. Now she would be able to attach a ventilator before bed and then remove it to breathe normally during the day. To this day, I still imagine her having to snap on a ventilator every night before she slips into sleep. In my craft, it's said that you mostly remember the ones you hurt. It's true.

Fortunately, the cyst in her brain never returned, so she would not have to face further dangerous surgeries. As I followed up with her over the ensuing months, I learned that some good had come of it. Although at first she took Lunesta to get to sleep and an anti-anxiety medicine to control her fears, she soon found that her conscious attention to breathing, all day, was actually having a calming effect on her.

"It's like I'm meditating all day," she told me.

And that, I think, makes for a segue to my next chapter.

7

JUST BREATHE

The first attack came a week after he learned that his parents were divorcing. Sitting in a class at his high school, worrying about the consequences, wondering where he would live, fourteen-year-old JT felt his heart pounding. He began to hyperventilate. It was a full-on panic attack. And then something strange happened: His left arm went dead. He couldn't lift it from his desk. Frightened, he went to stand up, but his left leg crumpled, and he collapsed onto the floor.

I heard about this a month later, when JT and his mother came to see me. He was a shy, quiet boy, thin and waifish. The kids in class had laughed at him that day, he said. But within twenty minutes, after he'd been carried by two teachers to the nurse's office, he had calmed down and recovered his strength. The nurse said he should go directly to the hospital, but he had refused. It was just some kind of freaky emotional response, he figured.

A few days later, talking with his father about the divorce, he had started worrying again, then hyperventilating and sweating—and then, for a second time, had lost all strength in the muscles on his left side. And, as before, his strength returned once he calmed.

Still, his parents didn't insist on taking him to a doctor. It's a funny thing about weakness. When somebody hurts badly, they want immediate evaluation and treatment. But weakness doesn't seem to catch people's attention as quickly, despite all the warnings about sudden muscle weakness often being due to stroke.

The final straw came a week later, JT's mother told me, when he had a second episode in school. He had been stressing again over the impending divorce, the worry had again evolved into a panic attack, and the muscles on his left side had gone flaccid.

This time the school nurse called an ambulance, which took him to the hospital, where the emergency room doctor conducted a physical exam and ordered an MRI brain scan. Everything looked normal on both the exam and the scan. No deficits in function. His central nervous system was "intact," as neurosurgeons like to say. No clots, no tumors, just beautiful juicy gray and white matter that filled the skull. A young, healthy brain.

But, obviously, *something* was wrong. The ER doctor called the radiologist, who astutely recognized that the MRI hadn't included a vascular study — a picture of the branching pattern of the brain's blood vessels. That required MR angiography, a special brain scan that reveals all the blood vessels, from giant vascular trunks to the smallest capillary branches.

Two large carotid arteries run up the neck on each side. Once they hit the back of your jaw, they dive deep, through gaps at the base of your skull, to enter the brain, where they fan out like the candlestick holders of a candelabra. In fact, we literally call it the candelabra — that's the anatomical term.

In JT, his candelabra of smaller vessels branching off from the left internal carotid was normal. But on the right side of the brain (the part that controls the left side of the body), there was a problem. The right carotid had somehow closed off, becoming a cul-de-sac rather than a tributary.

So why wasn't the right side of his brain dead? Because — like a tree that sprouts tiny new branches when a large limb is cut off — the stub of carotid artery at the base of his brain had sprouted a bushy network of miniscule vessels. Not nearly as thick and healthy as the candelabra of normal vessels, but just enough to keep his brain functioning under nonstressed conditions. On the angiogram, this fuzzball of thread-like vessels looked like a puff of smoke.

Seeing the image, the radiologist knew immediately what JT's problem was. The Japanese term for "puff of smoke" is *moyamoya*. Discovered and named by Japanese doctors in the late 1950s and early 1960s, moyamoya disease is most often diagnosed in children when the physiological stress of crying, panic, or exertion overwhelms their tiny vessels' ability to get enough blood to the brain. Usually, this results in bleeding of the thin vessels, leaking blood into the surrounding brain tissue, killing it. In other words, it's a temporary mini-stroke, also known as a transient ischemic attack. For some people, it causes headache, involuntary muscle contractions, repetitive twisting movements, or, as in the case of JT, loss of feeling and strength in the muscles controlled by that portion of the brain.

The ultimate fix for moyamoya is surgery to graft a new source of blood vessels into the area of the brain with diminished blood flow. That's why his doctors referred JT to me.

But JT's case was at an unusually early, mild stage. The image showed no evidence of the small dead patches of brain tissue that develop as the disease progresses. The constriction of his "puff-of-smoke" vessels had only temporarily robbed the nearby tissue of blood, without actually killing brain. So, for now, rather than risking surgery before it was absolutely necessary, JT could heal himself. The short-term solution was simple yet profound: He just needed to breathe.

Hyperventilation during a panic attack makes people feel like they don't have enough oxygen, but actually they have plenty. Their red blood cells, carrying oxygen, remain fully loaded: 100 percent saturated. But all that heavy breathing lowers the concentration of carbon dioxide — the gas we exhale. We actually need a certain amount of carbon dioxide circulating in our blood; if the carbon dioxide is too low, the brain responds by squeezing and shrinking the small blood vessels. For JT, the resulting decrease in blood flow would push him off the edge. Like a field of flowers that dry up in a drought, his thirsty neurons would falter and fail to fire. If re-irrigated with blood in time, they would spring back to life. Other organs can go hours without blood flow before their cells die, but neurons survive only a few minutes without blood.

And so I spoke with JT and his mother about doing his best to slow down his breathing whenever he felt a panic attack coming on, to calm his mind, and protect his brain.

JT's mother was skeptical, but he was all in. For months, JT was a diligent student, working through his anxiety by learning to breathe slowly. Controlling his breathing meant controlling his brain, which in turn kept his arms and legs working.

NEURO GEEK: MINDFUL BREATHING

While moyamoya is literally a one-in-a-million disease, mindful breathing can benefit anyone and everyone. First taught by Buddha over 2,500 years ago, it's a fundamental part of mindfulness meditation, focusing the mind on the here and now. Its benefits, however, are not just spiritual or psychological. Mindful breathing (or paced volitional breathing, as it's referred to in neuroscience) improves the very structure, physiology, and function of your brain.

One of the most extraordinary demonstrations of the effects of mindful breathing on the brain was described in a study in the journal

Neuroimage by German researchers based in Munich. They trained twenty-six people in mindful breathing for two weeks. Then they tested the participants' brain function inside an MRI machine. The participants were asked to either breathe normally or mindfully as they viewed disturbing, emotionally provocative photographs.

What they found was that during mindful breathing, connections were strengthened between the amygdala, an area where strong emotions (both positive and negative) are processed, and the prefrontal cortex, the brain's chief executive officer. This, the authors concluded, shows how mindful breathing helps the frontal lobe stifle negative emotions.

Other studies, by Michael Posner and Yi-Yuan Tang of the University of Oregon, have focused on a type of meditation called integrative body-mind training, which emphasizes mindful breathing. In one experiment, they found that eleven days of training in mindful breathing increased participants' white matter connections emanating from their brain's anterior cingulate cortex (ACC). Located just behind the frontal lobe, the ACC not only helps to regulate blood pressure and heart rate, but is also closely involved in decision-making, impulse control, and even ethics. In a second experiment, Posner and Tang found that just five hours of training over the course of two weeks increased neuronal branching in the ACC. An additional six hours of training resulted in those neurons getting insulated by myelin sheaths.

One last study before we get back to JT! It's a game changer. "Breathing Above the Brain Stem: Volitional Control and Attentional Modulation in Humans" was the title of a paper published in 2018 in the *Journal of Neurophysiology*. Researchers led by a brain surgeon from the Hofstra Northwell School of Medicine studied the pattern of electrical signals coming from different parts of the brain during regular breathing compared to carefully paced, mindful breathing, as measured by an EEG. This was not just the usual EEG that reads signals from the surface of the scalp but an invasive EEG. (The patients involved had epilepsy and agreed to have the monitors placed directly into their brain, while hospitalized, to find the source of the aberrant electrical discharges causing their seizures. During their extended hospitalizations for this monitoring, boredom is common. So many patients gladly participate in activities such as meditative breathing, providing us with direct measurements of associated brain activity.)

What the researchers found was that during slow, controlled breathing, signals coming from multiple areas of the brain were more in sync with each other than during regular breathing. They concluded: "Our findings imply a fundamental role of breathing-related oscillations in driving neuronal activity and provide insight into the neuronal mechanisms of interoceptive attention." (Interoception, sometimes considered a "sixth sense," is the perception of one's own body and its functioning.) This study is proof of the neurobiology underlying the mind-calming effects of paced volitional breathing, commonly known as mindful breathing.

BREATHTAKING

For JT, mindful breathing helped for a while but was no panacea. After nearly a year, his angiograms showed a few small patches of dying brain tissue, where the wispy "puff of smoke" vessels had leaked blood. There was now only one treatment that could help: he needed new plumbing, a new, larger network of blood vessels to feed the right side of his brain.

The operation I performed, called encephalo-duro-arterio-myo-synangiosis, is kind of bizarre. After shaving his head and peeling back his scalp, I had to release the temporalis muscle from where it connected to the upper jaw. This is the muscle underneath the stem of your sunglasses and above your ear, the one that visibly moves when you chew. Leaving in place the other end of the muscle, where it is anchored to the skull above the ear, I then slipped the loose end through a notch that I made in the skull. But here's where it gets weird: I simply left the slice of red, beefy muscle lying on the glistening white surface of his brain — inside the skull! — and closed the scalp. I didn't even connect blood vessels from the muscle to the blood vessels of the brain that so desperately needed them.

Why? I knew that over the next few months, seduced by vascular growth factors released from thirsty neurons, new blood vessels would sprout from that hunk of meat and grow straight into the portion of the brain that needed them. And so JT, who had cured himself of his panic attacks through mindful breathing, was cured of moyamoya through his brain's ability to lure blood vessels to where they were needed.

Surgeons call what I did "indirect revascularization."

I call what JT did breathtaking.

NEURO GYM:
HOW TO BREATHE MINDFULLY

Here is how to breathe mindfully: Sit down somewhere quiet, and pay attention to your breathing, putting other things out of your mind as best you can for ten or fifteen minutes.

Breathe in slowly, through your nose, for a count of four.

Hold your breath with lungs filled for a count of four.

Exhale slowly, through your mouth, for a count of four.

Wait to inhale for a count of four.

There. Easy, right? Well that's the breathing part. And it gets easier as you dedicate more time to it. But the mindfulness (being focused only on the activity at hand) part, where distractions are held at bay, is the harder part.

That's why many people like to attend a mindful meditation class or seek out a coach. If you want to get the most out of mindful breathing, I suggest you look around for a program in your area, where you can get personal instruction and support for a few sessions.

Just searching for a mindful breathing class is not going to get you far, though. Most offer it as part of a mindfulness meditation class. Yoga classes almost always incorporate mindful breathing as part of their program, as well.

There is another, easier, cheaper way. While the internet has no shortage of bad, inaccurate, or even dangerous medical information, you really can't go wrong with an app or YouTube video offering guided meditation and breathing. There are hundreds of free videos and dozens of apps. Classes and apps are a great way to start, but ultimately meditation is a deeply personal journey and one that is honed in solitude.

NEURO BUSTED:
THE BUSINESS OF MEDITATION

Eastern practices of meditation originated over 2,500 years ago, promoting the power of inward focus and breathing techniques as well as the renunciation of materialism. Today it's a multibillion-dollar industry. What has long been an intense and personal activity that offers physical and mental benefits is now packaged into a downloadable commodity with promises of digital nirvana. The experts are proliferating, since anyone with a brain can declare themselves a mindfulness expert.

My advice is to check out a few websites, maybe an app, or even try a class or two if resources permit. Learn the framework and get a few useful tips. Just don't set aside your instincts in the pursuit of mindfulness. Be wary of feeling pressured to purchase products or participate in expensive classes. In the end, what's in your mind is entirely your jurisdiction, so protect and nurture it. In its essence, meditation should reaffirm the maxim that some of the best things in life are free.

8
HOW TO HANDLE HEAD INJURIES

B ack in 2002, when I was a resident working at a trauma center in San Diego, some of the worst traumatic brain injuries (TBIs) happened when fools fired their guns into the sky at midnight on New Year's Eve. Those bullets came down like raining lead, injuring and killing people just as if they had been shot directly.

It's not as big of a problem as it used to be, thanks to local public service campaigns like "Bells, Not Bullets" and "Bullet-Free Sky." But back then, it was a totally crazy ritual that filled ambulances — and eventually nursing home — with victims.

Usually, when an EMS team picked up someone with a gunshot wound to the head, they called ahead and told us the person's condition en route to the hospital. This allowed our team of six — usually a trauma surgeon, an anesthesiologist, an X-ray technician, two nurses, and a specialized surgeon like myself — to be ready in the trauma bay.

But this particular night, a patient in his late thirties arrived on his own. At 12:30 a.m., Theo drove himself from the nearby beach community where he lived to our hospital. He walked up to the ER check-in counter and said he had been stargazing when he had been hit by what he thought was a rock. He said he felt fine and appeared calm. But the admitting clerk saw what looked like toothpaste — actually, it was the white matter of his frontal lobe — oozing from a hole the size of a dime high up on his forehead.

The clerk immediately called the nurses, who called our team. Theo was placed on a gurney and rushed to the trauma bay, where I met him.

The skin around the entry wound on his forehead was raw and torn, but I knew from the lack of burn marks that he hadn't been shot point-blank. Either he was shot from a distance or pierced from a falling bullet. Given his story, I was leaning toward the latter.

I ran through the three standard questions to check his mental status. He knew his name, the date, and where he was. There were no hints of frontal lobotomy–type personality changes. Theo was himself.

The back of his skull was still intact, no exit wound, so I had a few things to figure out. First, the dangerous stuff: where was the bullet and what was its trail of injury? They wheeled him in for a computed tomography (CT) scan to find out if there was ongoing bleeding.

The worst-case scenario would be a "talk and die," the kind of injury that led to the tragic death of actress Natasha Richardson in 2009, when she fell and hit her head while on the ski slopes in Canada. Reportedly lucid at first, when she finally arrived at a hospital by ambulance, six critical hours had elapsed since her fall, and she was soon declared brain dead.

I knew Theo didn't have a massive blood clot, yet. If he did, we wouldn't have been able to go over his personal details. I just didn't want to miss the one that would start small and blossom over the next few hours.

The scan showed us that the bullet had torn a narrow channel through his right brain. His soft gray and white matter offered no resistance as it tumbled, undeterred, and lodged into the bone encasing his brain, about four inches above the top of his neck. It also revealed two golf-ball-sized clots, one in his right frontal lobe behind the forehead, and the other in his right occipital lobe at the back of the head. If the torn vessels didn't seal themselves off, the blood clots would expand. In that scenario, I would have to take care of it, but that would require brain surgery. So we waited a few hours to see where things were headed.

Unfortunately, the next scan showed that both clots had increased in size. Still, Theo continued to engage me fully. I spoke to him about the precarious perch on which his brain teetered. The clots were hours away from sending him into a coma and possibly brain death. We discussed the risks, benefits, and alternatives to surgery. Even with the damage done, he was lucid enough to give his informed consent.

I called the operating room. "Trauma craniotomy," I said. Within minutes, a team of nurses, anesthesiologists, and technicians would be ready and waiting.

While being wheeled into the operating theater from intensive care, Theo sneezed — and brain sprayed out of his forehead. The sneeze had raised his intracranial pressure, and with the clots taking up ever more space, this was his brain trying to find a lower-pressure environment. Liquefied brain had exited its osseous cage into the atmosphere. This man was on the edge — he was about to talk and die.

Lying on his back on the operating table, fully anesthetized, Theo got the fastest haircut of his life. I quickly sliced his scalp down to bone. The electrocautery, used to singe flesh and vessels, was set on high, so high that flashes of small flames and sparks sprayed as my instrument discharged electricity onto the surface of the skull. Then I drilled ferociously to open his skull. The brain was tense and taut so I punctured the thin shell of cortex hiding the blood ball in his right frontal lobe with a large suction and pushed it hard directly into the middle of the clot.

Large gobbets of blood and ravaged brain were suctioned. Awful as it was, it had to be done. That part of his brain was irreparably damaged. Then, almost instantly, his frontal lobe deflated, like a soufflé falling flat. No longer did it press up against the inside of the skull.

Now came the time for finesse. I took particular care with the delicate boundary between angry red clot and pearly white brain. Get too obsessive about removing every last smudge of congealed blood, and you risk losing another million brain cells.

Once the crater where the clot had been emptied, I cauterized shut the surrounding blood vessels and filled the cavern in his brain with sterile water.

"Valsalva," I said to the anesthesiologist.

With this instruction, he manipulated Theo's ventilator to recreate a sneeze. Under this increased intracranial pressure, my micro-welding of the blood vessels was put to the test. And then: nothing. No faint swirls in the crystal water. No blood. So I closed up this section of the brain and skull. I also addressed the entry point on his forehead by leaving just a small piece of wire mesh over the dime-sized hole left by the bullet, with only the skin covering it.

Night had turned to dawn, but another lobe still required evacuation: the occipital. Both the clot and the bullet were lodged up against the back of Theo's skull, so we turned him over onto his stomach. I removed a three-inch by three-inch blunted square of the skull, and there it was: the bullet, wedged right into the bone. The skull there was about the thickness of a forefinger, thicker than elsewhere. Using a surgical chisel, I chipped out the bullet, suctioned out the clot underneath, then closed it up. He went home six days later.

Until spending the first half of New Year's Day operating on Theo, I never imagined someone could survive a gunshot wound like he had suffered without colossal damage to his mind. But months later, Theo drove himself back to my clinic. On scans, parts of his brain were still missing. They didn't grow back—they never will—but the remaining brain, both fragile and resilient, continued to suffice. Most importantly, Theo was doing fine. He had a small loss of peripheral vision on his left side — nothing else — and otherwise made a complete recovery. Theo went on with his life having survived a bullet to the brain.

Was he lucky? Definitely. But years later, the first study ever to survey the outcome of patients like Theo found that Theo's experience wasn't all that rare: 42 percent of people shot in the head survive and do well enough to be discharged from the hospital within six months. I doubt that many of them do as well as Theo, but still, even as a brain surgeon, I find that incredible.

Of course, you are far, far less likely to get shot in the head than you are to suffer a concussion or other head injury. Every year, according to the Centers for Disease Control and Prevention, nearly 2.8 million people in the United States end up in the hospital due to a traumatic brain injury. Almost half of those are due to falls; about one-third result from automobile accidents. But increasing concern has arisen in recent years about concussions incurred while playing sports — particularly when the players have a second concussion before they have recovered from the first.

NEURO GEEK: WHEN SYMPTOMS PERSIST, IS IT "ALL IN YOUR HEAD"?

It is true that, in rare cases, the effects of a concussion can last for months before finally resolving. Even more rarely, and for reasons that continue to puzzle researchers, some people continue to report mood swings, concentration problems, fatigue, dizziness, and other problems that go on for years.

Some doctors, unfortunately, can be dismissive of such symptoms. "It's all in your head," they might say. Or, "I can't find anything wrong with you, so I can't help you." But studies have shown that most such people are truly suffering, so what's going on?

The most enlightened approach to helping these patients with long-term symptoms is to understand that they have a *functional* illness, one that may or may not involve any permanent, observable injury to the brain but which nevertheless changes how the brain functions.

"That doesn't mean that people who have dizziness or these other symptoms where we cannot find a specific cause are making it up," says Terry Fife, MD, of the Barrow Neurological Institute at Dignity Health St. Joseph's Hospital and Medical Center in Phoenix. "They may be severely disabled by it. And it sometimes responds to medication." Usually, he says, those medications are antidepressants.

Jon Stone, PhD, who specializes in functional disorders at the Center for Clinical Brain Sciences at the University of Edinburgh in the United Kingdom, says that in most cases the disorder is triggered by some kind of injury. "But then instead of recovering and getting well," he says, "the person gets stuck with the same symptoms persisting over time."

If you or a loved one continues to suffer from the effects of a seemingly mild head injury for weeks or months, seek help from a doctor or clinic specializing in brain injuries and concussions. If no such center is located in your area, seek out a physical or occupational therapist who is willing to help you get moving, slowly, again. If you're struggling with feelings of depression or mental confusion, consider reaching out to a psychologist or other mental-health professional.

THE RISKS OF MULTIPLE CONCUSSIONS

Multiple blows to the head, whether in football, soccer, boxing, hockey, or other activity, can sometimes cause permanent brain damage accompanied by changes in personality, memory, and thinking.

Chronic traumatic encephalopathy, or CTE, is the medical term for the damage caused by these repeated concussions. Standard MRIs cannot detect it, so a diagnosis of CTE usually cannot be made with certainty until after a person has died. Only on autopsy can a pathologist see the damaged brain tissue.

Boston University's CTE Center has established a "brain bank," where former athletes with symptoms consistent with CTE can donate their brains upon death. Now with 425 brains, the center published a study in 2017 of former professional and amateur football players. Among 111 NFL players, the brains of all but one of them showed signs of severe CTE.

That's sobering and frightening, but it's extremely important to keep in mind that this study involved former players who already showed personality and mental changes consistent with brain injury. It was not a random sample of NFL players, most of whom never show such changes despite having experienced concussions.

The same study saw far less widespread damage to the brains of football players who never reached the NFL. Only three of fourteen former high school players had CTE when their brains were examined, even though their brains had been donated specifically because they or their families thought they might have CTE.

Another study, by the U.S. Centers for Disease Control and Prevention (CDC), gave a far more representative picture of how former NFL players fare. Among 3,439 who had played at least five seasons between 1959 and 1988, the death rate from all causes was actually about *half* that of the U.S. average. Only 334 of the retired players had died by 2007, and of them, just 2 had died of dementia. By comparison, 85 had died of cancer and 126 of heart disease.

CTE is a terrible thing; it ruins minds and destroys lives. But recent media reports have created a false impression that every football player — and anyone else who experiences concussions — will suffer inevitable brain damage. That's just not true.

The exaggerations over CTE can have tragic results. In 2016, former NHL player Todd Ewen became convinced that he had CTE. At the age of forty-six, having suffered with depression for several years, he committed suicide.

"Every time it was announced that a fellow player had CTE, Todd would say, 'If they had CTE, I know I have CTE,'" said his widow, Kelli, in a statement released by the Canadian Concussion Centre. "He was terrified by the thought of a future living with a degenerative disease that could rob him of his quality of life, and cause him to be a burden to his family."

And yet, when examined at autopsy, Ewen's brain showed no physical signs of brain damage associated with CTE.

NEURO BUSTED: THREE FALSE BELIEFS ABOUT CONCUSSIONS

There are three common misunderstandings about what a concussion is and what it is not.

If you trip on a rug and hit your head on the edge of a table, getting a small cut on your forehead in the process, does that mean you have had a concussion?

Not necessarily so. Whether or not you bleed or have any external signs of head injury has nothing to do with a diagnosis of concussion. In fact, even an MRI or CT scan of your brain usually shows nothing wrong.

The second point of confusion among the public is the mistaken idea that a concussion requires temporary loss of consciousness. Not true! In many cases, the person remains wide awake immediately following a concussion.

So how *do* you diagnose a concussion if neither physical injury nor loss of consciousness is required?

The one and only factor in diagnosing a concussion is simple: It *must* include a change in mental functioning either immediately or in the hours after a blow to the head. The person might feel dizzy, confused, or nauseated; he or she might develop a headache. They might have temporary trouble talking, walking, remembering, thinking straight, making decisions, or doing anything that requires muscle coordination. Their eyes might suddenly be sensitive to light. They might vomit, hear a ringing in their ears, or have visual disturbances.

Of course, most people do not die following a blow to the head, and that leads us to the third most common misunderstanding about concussions: The well-deserved media attention to the plight of professional athletes who developed lifelong disabilities due to *multiple* concussions has led many people to think that even a *single* concussion results in permanent harm. In reality, the vast majority of concussions leave no lasting effect on a person's mental functioning. In a matter of days or weeks, the person feels fine and shows no mental or emotional deficits.

THE TRUE RISKS OF FOOTBALL AND OTHER SPORTS

Publicity about CTE among retired NFL players has become so intense that many parents now forbid their children from playing football. Dr. Bennet Omalu, the pathologist who first described the disease, has gone so far as to say that permitting a child under the age of eighteen to play football is "the definition of child abuse." Personally, I consider that view to be over the top. Soccer, after all, may be even more likely to result in concussions than football. A study by researchers at McGill University in Montreal found that 46 percent of their soccer players had experienced a concussion during a single season — more than the 34 percent of football players. And a majority of both soccer and football players who had one concussion had a *second* concussion during the same season.

That second concussion is the one to worry about. Whereas the vast majority of people recover from a single concussion with no long-term damage, those who experience a second concussion before recovering from the first are at increased risk of developing CTE. After all, that's why it's called *chronic* traumatic encephalopathy — because it's the result of multiple concussions.

In deciding whether to let your child participate in contact sports, you deserve to know the truth. No one ever suggested that boxing was safe for the brain, and now we know that other contact sports can be similarly dangerous to your brain and mind.

If you want to let your son or daughter play football, hockey, or soccer, that's your choice as a parent. But if they experience a concussion, I suggest you urge them to consider switching sports.

NEURO GYM: HOW BEST TO RECOVER FROM A CONCUSSION

The first rule of recovery from a concussion should be obvious by now: Don't run out and get a second concussion. Especially until after all symptoms of an initial concussion have cleared up, it is absolutely essential that a person avoid activities that could result in a second one. A quick second concussion increases the risk of developing long-term effects.

Translation: Don't send them back in the game, coach!

But beyond that piece of common sense, how much rest does a person really need after a concussion?

Some medical societies and hospitals now recommend what is called "complete" rest. You will see this written on dozens of health websites. They say a person should avoid all physical *and* mental activities for a week or more. They insist that something called "cognitive rest" is necessary. This means no reading, homework, job activities, video games, texting, email, web surfing, or social media. They even insist a person should wear sunglasses and stay in a dark room, lying in bed, literally doing nothing. It's called cocoon therapy, and it's total science fiction. In fact, it's harmful.

The brain requires stimulation to function normally. We know from decades of research that an "enriched" environment enhances brain development in people and animals alike and that an "impoverished" environment kills brain cells and leads to lifelong cognitive deficits in children. So, it shouldn't surprise anyone that putting a kid who hit his head into the equivalent of a dungeon for a week might not be the best of treatments.

A series of studies have found that reaching a middle ground produces the best results: young people do best following a concussion when they avoid doing anything in which another injury might occur but otherwise engage in moderate physical and cognitive activity. For instance, a study published in the journal *Pediatrics* in 2015 was actually designed to show that strict rest would benefit people between the ages of eleven and twenty-two following a concussion. Instead, the researchers ended up finding the opposite: participants who had been assigned to strict rest had *more* post-concussive symptoms during the ten days after their concussion than those who were told to simply take it easy.

So if you or a loved one has had a concussion:

1. See a doctor.
2. Take it easy for at least a few days.
3. Avoid cocooning yourself. A little texting is not going to hurt you.

9

FOOD FOR THOUGHT

E levators are for the patients; doctors take the stairs."
That phrase has stuck with me ever since Professor Demetri
said it at Romodanov Neurosurgery Institute of the State Institution
in Kiev, Ukraine.

It was the summer of 2004, and I had just arrived at the hospital
directly from the airport, where he had waited for me holding a sign
with my name written in Ukrainian. I had emailed neurosurgeons
in developing countries around the world, offering to collaborate on
advanced techniques for minimally invasive neurosurgery. Demetri
had been among the first to respond, so here I was, having flown to
Ukraine during a vacation week from my residency in San Diego.

The outside of the hospital looked like a battleship, with years
of wear and tear. It was a government hospital, which meant it was
poorly funded and used only by the poor. Open windows dotting the
austere weather-beaten concrete exterior had gossamer, sheer curtains billowing out. Like a tank with lace.

The neurosurgery institute was on the top floor, so once inside the main entrance, I went to press the button beside the only elevator door.

"Elevators are for the patients," Demetri said. "Doctors take the stairs."

He led me through an open door to the stairwell. We began climbing.

On the first floor, the fire door to the hallway was open.

"You don't keep these closed?" I asked.

"No air-conditioning. It gives breeze. If fire comes —" He arched an eyebrow, shrugged, and kept climbing.

The stairwell was bleakly evocative. On each floor, the sharp smell of bleach and soap floated out the doorway. But each floor also had a distinctive smell that couldn't be veiled. The smell of amniotic fluid on the delivery ward. The antiseptic tang of the ICU with its machines pinging in the distance. The lingering smoke from the cauterization of human flesh — the scent, so familiar to me, of operating rooms. And then, on the floor for patients with brain diseases, a scent that was completely unfamiliar: like bacon cooking in a frying pan mixed with cleaning fluid, kerosene, and the medicinal smells of a pharmacy — an olfactory olio.

I gave Demetri a quizzical look, and he led me down a hallway into the first room on the right. Six beds filled the room, with six children lying in them, some sleeping, some sitting up. They were so thin, I thought at first they were oncology patients. Two of the mothers were lying beside the children; the others were talking or cooking on little kerosene burners in the corner. One of the burners had a pot of water boiling, with metal needles and glass syringes being sterilized. Another had something sizzling in a pan.

I asked Demetri what it was.

"Pig fat," he said.

And then I understood. Silently, I mouthed the word to Demetri: "Epilepsy?" He nodded.

I should have known: The kids were being fed a diet of fat, fat, and more fat. Only fat. More than a hundred years ago, physicians discovered that a diet consisting almost entirely of cream, oil, butter, and other fats could greatly decrease or even eliminate seizures in children with epilepsy. The treatment had been pushed aside when the first medicine that could reduce seizures — phenobarbital — was discovered around the same time. But the diet made a comeback beginning in the 1990s as a treatment for when anti-epileptic medicines fail.

In Ukraine, Russia, and other former Soviet states, however, the diet had never gone out of favor, particularly among families who could not afford anti-epileptic medicines. And so the mothers were making their own medicine, crispy-fried pieces of pork fat, to settle down their children's aberrant electrical brainstorms.

NEURO BUSTED: YOUR GUTS ARE NOT A SECOND BRAIN

Nearly every millimeter of your body is penetrated by nerves that have been sent out from the brain. One particular nerve net covers your stomach and intestines and is a well-known anatomical structure called the enteric nervous system, or ENS. The ENS is how you feel butterflies in your stomach when your mind is anxious. It also carries signals of hunger and satiety up to your brain. Are these nerves in the gut important? Absolutely. Have they been given more import than they have earned? Absolutely. Abdominal surgery provides the key insight to why the role of the enteric system has been inaccurately hyped lately.

For several medical conditions (bowel obstruction, cancer, intra-abdominal adhesions), large lengths of bowel can be removed. Surprisingly, seventeen of the twenty inches of the small intestine can be removed with little functional consequence. Most of the colon can be removed as well. In fact, nearly the entirety of the intestines can be removed. Yet remarkably, there are no reported neuropsychiatric effects in these patients after decades of experience around the world. Nerves around our guts? Yes. Worthy of being called a second brain? Not from what I have read and seen.

THE KETOGENIC DIET

The original diet for children with epilepsy, developed early in the twentieth century, was designed to deliver over 90 percent of calories by fat, with very little protein and almost no carbohydrate. Why it worked to reduce seizures was unknown when the diet was first developed and remains something of a mystery to this day.

But this much we do understand: With a normal diet, neurons and other cells use glucose, a sugar derived from carbs, as fuel. But without carbs, or when you use up your carbs in storage (typically in about sixteen hours), the liver begins turning fat into ketones, a backup energy source. And for reasons that we have yet to fully understand, neurons that get their energy from ketones instead of from glucose become less excitable, less prone to firing out of control and causing seizures. And so ketones from fat act as medicine for these children.

Once anti-epileptic drugs became available in the 1950s, the ketogenic diet fell out of favor in Western countries as a treatment for epilepsy. But in the late 1960s, a doctor named Robert Atkins decided to try a modified ketogenic diet in an attempt to lose weight. The book he published in 1972, *Dr. Atkins' Diet Revolution,* became a huge bestseller, despite the fact that the medical establishment was then convinced that fat was an evil to be avoided at all costs. The diet he developed, and which some people still follow, includes not only unlimited amounts of fat, as with the strict ketogenic diet, but also unlimited amounts of protein and low-carb vegetables.

I support sharply limiting carbs and the emphasis on eating plants. But unlimited fat and protein is not something I can recommend because you're trading one problem for another: too many calories and bad cholesterol. Let's dive into what you should eat if peak brain performance and healthy aging is what you seek.

THE MIND DIET

The Mediterranean-DASH Intervention for Neurodegenerative Delay diet — or MIND diet, for short — was specially designed to improve brain health. Recent well-done studies have found that sticking to the MIND diet helps people avoid mental decline and remain cognitively healthy. One study even showed that people who stuck to the MIND diet cut their risk of developing Alzheimer's disease *in half*. That's extraordinary. And since no drug has yet been developed to prevent dementia, it's your only move.

My family tries to follow the MIND diet but not too strictly. It's big on fresh fruits, vegetables, nuts, fish, and chicken, and down on red meat, saturated fats, and sweets. It is simple and delicious, but I want to emphasize that we use it as only a basic guideline. We do sometimes eat steak or bliss out on chocolate. More information about the MIND diet, in detail, can be found at https://www.webmd.com/alzheimers/features/mind-diet-alzheimers-disease.

Of course, the MIND diet is extremely different than the ketogenic diet that the kids in Kiev had to follow out of necessity. Thankfully, eating all that fat is not the ideal approach for most of us. But the brain benefits of ketones can still be yours through another route when you complement your MIND diet with intermittent fasting.

INTERMITTENT FASTING

Perhaps there is a reason why most of the world's major religions call for periodic fasting. Intermittent hunger clears the mind, awakens the senses, and improves brain functioning. Plus it lowers your blood sugar, reduces your insulin levels, and helps you lose weight by reducing total calories. What's not to love?

Well, the hunger. But it only lasts for a short time!

Consider our prehistoric ancestors, the hunters and gatherers who survived through feast and famine, abundance and scarcity. The *real* "Paleo diet" didn't consist of just large hunks of meat. Many were the days and weeks they failed to catch an auroch or boar and went to sleep hungry.

But with the hunger pangs come benefits. Going without food for even a day increases your brain's natural growth factors, which support the survival and growth of neurons. Evolution designed our bodies and brains to perform at their peak as hybrid vehicles. Metabolic switching between glucose and ketones is when cognition is best and degenerative diseases are kept at bay. As a recent paper in *Nature Reviews Neuroscience* put it: "Metabolic switching impacts multiple signaling pathways that promote neuroplasticity and resistance of the brain to injury and disease."

So how do you do it? Not by overloading on glucose or ketones, but by altering the cadence of eating and letting the body do what it was designed to do during times of food scarcity.

I'm not talking about caloric restriction, which extends longevity in animals and may well do the same in humans. People who follow a serious caloric restriction diet, eating as little as a thousand calories per day, are *always* hungry. I'm talking about being intermittently hungry by forcing your body to burn its fat reserves once or twice a week. The exhaust from this, ketones, will not only keep your brain going during those periods of fasting and hunger but will actually improve cognition, grow the connections between neurons, and stave off neurodegeneration.

I follow (or at least try to) an intermittent fasting diet, and I recommend it for anyone who wants to improve their mood and hit peak cognition. Here is my plan:

FAST TWICE A WEEK. The goal is to hit two stretches of sixteen-hour periods without food. So choose two days, not back to back, and skip breakfast and lunch or lunch and dinner. When you add in the hours you sleep, it's relatively easy to get to sixteen hours. Every Monday and Thursday, I skip breakfast and lunch and eat only dinner. Whatever my wife and sons are having, that's what I have.

NO BREAKFAST. I'm not talking about just on fasting days; I'm talking about avoiding breakfast almost *every* day! Some people insist that breakfast is the most important meal of the day, but there's no good evidence for that. The only time I eat breakfast is occasionally on the weekends, with my boys, just to hang out and be in the moment with them.

SALAD FOR LUNCH. I rarely eat a sandwich or burger or anything with carbs. My routine is to have a salad for lunch. It's a little painful.

NO LATE-NIGHT SNACKS. This one is hard for me, especially after a long day or when I have fasted. But I try.

Please keep in mind, I'm no extremist. I do go out to eat with family and friends, often. Sometimes I'm invited to a breakfast meeting and go with the flow. But I have made intermittent fasting part of my routine.

On days when I am operating, in fact, I eat nothing until late afternoon. I don't even have a cup of coffee, because once I enter the OR, there is no skipping out to the bathroom. I am routinely in there working for eight hours straight without a break. It may sound surprising that I'm not dragging from lack of food, but quite the opposite: I find it keeps me more alert.

NEURO GEEK: YOU ARE *NOT* WHAT YOU EAT

Most of what you eat will never make it into your brain because of the blood-brain barrier. When the arteries from your heart penetrate the skull and become brain arteries, they are no longer as porous as they were outside the brain. Instead, they become lined with a thicket of specialized cell layers, sharply limiting what can cross from the bloodstream into your brain tissue. The discovery of the nearly impervious blood-brain barrier occurred in the late nineteenth century with a simple experiment. Blue dye was injected into the blood of a mouse and then an autopsy was performed. The entire body was blue except the brain and spinal cord, which remained white. The dye couldn't penetrate.

We now know that even inflammatory cells and most of the medicines that work elsewhere in the body cannot cross into the brain. This makes it especially difficult to develop drugs to treat neurological problems, as I learned when working on treatments for brain cancer.

What does get across? Mostly just oxygen, glucose, and ketones. Some fats, vitamins, and minerals get through as well. Nearly everything else the brain needs it builds in-house. So when you hear about "brain food," keep in mind that the brain is a very picky eater.

WHEN DIET ISN'T ENOUGH

Most of us hate to take prescription medicine. Diet and exercise is how we all want to lower our cholesterol levels, lose weight, lower our blood sugar level, and cure ourselves of diabetes, depression, and other common ailments.

Some quacks try to exploit this natural preference by claiming that diet, exercise, and natural supplements are the *only* safe way to treat diabetes. That's a dangerous lie.

J.I. Rodale — the founding publisher of *Prevention* magazine, advocate of health foods, and critic of mainstream medicine — famously said on a television show that he would live to be 100. He then promptly died right there on the set of the show, sitting next to host Dick Cavett, at the age of 72. (The broadcast, alas, was never aired.)

As for Robert Atkins, he also died at age 72.

Diet, in other words, is not all powerful. So when diet and exercise aren't enough to help you stay healthy, please think about the medicines that can work safely and effectively.

The first-line treatment for type 2 diabetes, one of the oldest and cheapest medicines for the disease, is metformin. Taken as a pill, metformin not only makes your muscles more sensitive to insulin so that your body produces less of it, but also lowers the amount of glucose your liver drips into the bloodstream. Best of all, metformin has been shown to improve the long-term cognitive functioning of people with diabetes. Studies have shown that it can reduce the risk of developing dementia by up to one-third. And it is the *only* diabetes medicine that was found in an Australian study of older adults to protect against the loss of memory, executive functioning, and verbal learning. I'm not saying you should take this medicine to boost brain power; I'm just saying that several approaches can be taken into consideration.

NEURO GYM: LET YOUR HABITS HELP YOU

You have a busy life. Are you going to check an app, website, or booklet every time you want to get a bite?

Me neither.

The real key to improving your brain health is establishing basic, sensible habits that will get you through *most* days. Habits are powerful.

But as everyone knows, establishing new habits is hellishly

difficult. Very few New Year's resolutions last till summer, let alone till the next New Year's Eve. That's why it's important to be very selective and strategic in how you go about forming a new habit. First, be specific and positive: make it measurable as something you will do not just something you *won't*. Second, tell other people what you're going to do and ask for their support. Third, pick only one habit that you're going to change.

Once your habit is established, it becomes much easier. At my home, meals are about getting together and being with each other . . . not about whether we ate more than four ounces of cheese or forgot to serve kale.

We don't fuss too much. My wife and I have made a conscious effort to avoid turning food into a matter of guilt and stress for our boys. When I say that we follow the MIND diet, what I really mean is that we emphasize fresh, wholesome foods and avoid junk *as a routine habit*.

Given that many people in the world do not have access to fresh, affordable food, we try not to forget how lucky we are. This perspective is only reinforced the more I travel to developing countries, usually with one of my sons now that they're older.

I love the simple recommendation that author Michael Pollan has made: "Eat food. Not too much. Mostly plants."

That's it!

I tell my boys that fresh vegetables, fruits, nuts, and fish are the mainstays. Let everything else be an indulgence. Burgers: once in a while, not daily. Cheesecake: once in a while, not every day. Enjoying a meal and conversation, daily.

It's the habits that matter, I tell them, not the indulgences.

10
HOW THE BRAIN HEALS ITSELF

J ennifer was six years old when the symptoms started, so subtle that her parents thought they were a quirk of a young girl's development. Sudden feelings of worry and fear would prompt her to run into their arms. A few months later, though, the feelings became more intense and were accompanied by visions of seeing strangers who weren't there. Her parents took her to the doctor, who said it was nothing to worry about. But then the bouts increased from weekly to daily to several times a day.

One day Jennifer passed out and fell to the floor. Seconds later she awoke with no memory of what had happened. Now the doctors knew what to do. Jennifer needed to have her brain scanned. A tumor could have caused the seizure that stole her consciousness for those seconds.

The MRI revealed her brain to be structurally immaculate. All the ridges, the iconic curves, the net of blood vessels, the fluid chambers, the bone — everything was textbook normal. And so the search continued. Electrodes were placed over her scalp in search of aberrant electrical signals. Again, normal.

For the next step in her workup, she was admitted to the hospital for several days of 24/7 brain-wave monitoring with an EEG. One morning, while eating, it struck again: an intense bout of fear. At that moment, the normally smooth, rhythmic lines tracking her brain waves flickered and raged. A spark of epilepsy. That brief feeling of fear was her temporal lobe's warning of a convulsion that hadn't quite come, known as an aura.

Her parents were confused at the diagnosis. Didn't epileptics fall to the ground with their limbs jerking? But Jennifer's epilepsy was less severe for now, and the medicines the doctors prescribed allowed normalcy to return.

After a few months, however, the fits of fear returned. A higher dose was prescribed and a second medicine added. In a year's time she was up to three medicines and feeling groggy, with the spells of fear still breaking through. And then, on a visit to the epilepsy clinic, Jennifer passed out again. This time her brain declared that it would not be quelled by simple medicines. The brain arrhythmia, a seizure, spread from her temporal lobe to surrounding regions. The lightning storm swarmed into the entire right hemisphere of her brain, escaped across to the opposite left hemisphere, and then ricocheted back and forth. It was a grand mal seizure.

The clinic was attached to a hospital, so she was taken to the pediatric ICU. The strongest sedatives were squeezed into her veins to break the seizure, but that required so much medicine that she was essentially under general anesthesia. Her brain's urge to breathe was so sedated that she was attached to a breathing machine.

She was in a double bind: Left uncontrolled, continuous seizures would damage the very brain tissue from which they arose; but controlling the seizures would mean life in an unconscious state from sedation.

There was a third option. That was why Jennifer and her family came to us, flown with a team of medical attendants. When I met them, her parents' eyes were hollow. They knew by now that the usual kind of brain surgery performed for intractable cases of epilepsy was impossible for their daughter. Jennifer's electrical spikes did not emerge from any one spot that could be safely removed, like a tumor, but arose randomly from across the right hemisphere of her brain.

The medical teams had warned the parents about the next step in this tragedy. Now it was time for me to speak of the unthinkable with her parents: the only hope for their daughter, I told them, was to remove half her brain.

NEURO BUSTED: YOUR BRAIN IS *NOT* HARDWIRED

That mysterious organ in our skulls has often been explained by metaphors that seem inept in retrospect. The ancients believed there was phlegm inside the brain, while others suggested animal spirits permeated its recesses. More recently, the industrial revolution made gears the favorite explanation of the inner workings of our minds. Naturally, in our digital age, "wiring" has become a popular metaphor.

But neurons are multidimensional in the roles they can play to serve the brain's evolving needs. Ultimately, no neuron is hardwired to perform one task and can, to varying degrees, assume new functionality that previously was not expected of it. So, is there a better way to conceptualize the brain?

In neuroscience, we describe the brain in terms of "neuronal ensembles" that coordinate at a functional level. As the story of my patient Jennifer illustrates, when the physical structure of the brain is manipulated, or even removed, the remaining members of the neuronal orchestra can still work in concert to produce an amazing symphony of thought, imagination, and emotion.

INDISPUTABLE PLASTICITY

No example of the brain's capacity for self-reinvention is better demonstrated than in how it responds to having its left or right half removed in an operation called a hemispherectomy. First performed on a human in 1923, hemispherectomy is the most radical and outrageous surgery imaginable. Ignored for decades as being too dangerous, the operation was resurrected and improved upon beginning in the 1970s by surgeons at Johns Hopkins. Since then, it has become, if not common, at least not a last-ditch gamble. Not only do 96 percent of children experience a total or significant reduction in epileptic seizures following it, but few show any significant effects on their memory, intelligence, personality, or even sense of humor.

In neuroscience, this capacity of the brain to reassign and reinvent is called "plasticity." Through the 1980s, scientists believed that most areas of the brain were permanently restricted to handle particular tasks. Remember the picture of the somatosensory map shown in chapter 1, where the brain's areas for processing the sense of touch are assigned to specific spots for the cheek, the tongue, the fingers? Once those spots were mapped, everyone assumed they were permanent geography, like a map of North America.

But then along came experimentalists like Bradley Schlaggar, a young researcher at Washington University in St. Louis. Working with embryonic and newborn rats, he sliced out a portion of their visual cortex and placed it in their somatosensory cortex — in a spot where they perceive their whiskers. In this whisker-feeling area, neurons separate into darker areas that look like miniature barrels and lighter areas surrounding them. Each dark barrel is devoted to perceiving a single whisker. That's right, they're whisker barrels.

Before the results of Schlaggar's experiment were published on the cover of the journal *Science*, it was assumed that this structure was foreordained by genetics. But what he saw after placing neurons from the visual cortex into this area was astonishing: after a matter of weeks, these neurons that had been slated for vision organized themselves into the usual array of barrels for perceiving whiskers.

Around the same time, Michael Merzenich at the University of California, San Francisco, disabled one or two fingers of laboratory animals and then waited to see how their brains responded. Over a period of months, he found, the brain area that had been devoted to sensing the disabled finger was taken over by the remaining fingers. And with the additional brain space, the remaining fingers grew increasingly sensitive so that they could sense finer and finer pinpricks of stimulation.

Today we understand that plasticity is what the brain *does*, and not just when scientists do crazy experiments. But plasticity is not without limits, or at least we have yet to learn how to push past those limits. And plasticity is a double-edged sword with people suffering from post-traumatic stress disorder, who, for instance, continue to be haunted by intensely disabling fear, stress, and vivid recollections for years following a triggering event.

Following hemispherectomy, some kids never regain full control of the opposite side of their body. Surgeries to remove the left half can be especially concerning because that is where Broca's area and Wernicke's area lie, on the banks of the sylvian fissure that separates the temporal lobe from the frontal and parietal lobes. Broca's area enables the use of speech, while Wernicke's controls the *understanding* of speech. Lose them in adulthood, and the results are tragic. In very young children, the remaining right half of the brain will usually develop the capacity to speak and understand language, but rarely as well as with an intact left brain.

Jennifer, however, was lucky — if a child in need of a hemispherectomy can ever be so considered. Her left brain was fine. It was her right brain where the electrical signaling had gone haywire. That was the section that would need to be removed to give Jennifer's healthy left brain a chance to function unimpeded — and Jennifer a chance to grow and live.

BRAIN AMPUTATION

Through soft hair, I felt for her skull. Long, wide strokes of the clipper, and her hair fell. I painted betadine over her shorn scalp. The rust-orange liquid dripped off her neck and face. Under the intense beam of my head light, smoke and bone dust soared like embers off a flame. It had an unforgettable smell. Wood chips and smoke and something more. Usually small openings are made in the skull, but this time the sheer size of bone was closer to removing a continent off a planet.

An operative microscope descended from the ceiling and floated above Jennifer's brain at just the right level for my face to peer through its eyepieces. Odd as it may sound, a mouthpiece (shaped like a boxer's bite block) juts out below the eyepiece. With it clenched in my teeth, I twisted and rotated my head, and the microscope followed along, changing my field of view so that I need not take my hands from the surgical field. I neglected the painful spasms in my neck and performed the task at hand.

My left hand held a suction tube, with a flat area on top where my thumb rested. Sliding my thumb backward, I let air escape to lessen the suction; sliding forward increased it. My right hand held an eight-inch electric cauterizing forceps, which I could switch on or off with a foot pedal to singe shut small blood vessels.

I have stood in this place hundreds, thousands of times. The gear is so familiar, it is like an extension of my own body. As usual, with each heartbeat, the brain below throbbed softly. It showed no stigmata of disease. No dark color of blood from trauma; no cluster of mangled tissue of a cancer. The ridges and mysterious recesses displayed brilliant hues of red and blue microvasculature, mesmerizing speckling like a Jackson Pollock painting. It looked as it should. It looked perfect, normal, healthy. And I was about to tear it asunder.

It was slow work. Brain tissue isn't cut like other tissue; it's delicately chiseled by the force of suction. I started in on the right frontal lobe, first separating it from the temporal by suctioning a narrow trough at the bottom of the fissure that runs between them. Then I freed it from the falx, the brain's central keel of thick and doubled dura that divides the two hemispheres, protecting the corpus callosum below.

As the metal tip of my suction tube moved along, caressing away brain tissue, I felt an occasional bounce from bumping into a blood vessel, like lightly plucking a string on a guitar. Each time, I moved in with the forceps to cauterize the vessel. Then, without a word, my scrub nurse, standing to my right, replaced the cauterizing forceps in my right hand with a spring-loaded micro-scissor. She knew the sequence of instruments I needed; and at my shoulder, she knew that when my breathing shallowed I was about to take a risk. With her at my side, I never looked away from my canvas. I severed a singed vessel; she took the scissors and replaced the cauterizing forceps. Hundreds of times we repeated this maneuver in rhythm, without a false move.

As I circumnavigated the frontal lobe, systematically dividing the fine vasculature, its heavenly iridescence faded and darkened. Finally, nothing held it in place anymore. The scrub nurse took the forceps and suction tube from my hands and placed a spatula in each. With them, I lifted Jennifer's right frontal lobe, as though lifting an omelet, and let it slide off the spatulas into a gray metal basin.

Now for the parietal lobe. I removed the vessels and fibers connecting it to the temporal lobe on the right and the occipital behind. The parietal holds the ridge of neurons controlling all movements on Jennifer's left side. After this dissection, the steering wheel for the left side of her body was gone. I cauterized and severed the final blood vessels and then lifted and placed the lobe in another cold steel basin.

After the occipital and temporal lobes were removed, it was time to go deeper, to suction out the white matter, the right hippocampus, the right amygdala, the right thalamus and hypothalamus — right down to the brain stem, the basement, where I stopped.

To complete the operation, I ensured that the glistening corpus callosum that linked the two hemispheres had been properly sealed off. Toward its front is a section called the genu. Bent down and back, it is easily separated. Toward the back of the corpus callosum, however, the splenium demands respect: immediately beneath it lies the vein of Galen, one of the largest, deepest veins in the brain, named after the ancient Greek physician who discovered it. With the corpus callosum completely disconnected and sealed off, I saw what few have seen, what few *should* see: the great vein of Galen in a living human.

The remaining half of her brain was now her whole life. It was late afternoon when I finished. We decided to leave her on a breathing machine through the night. Driving home, I struggled to shake off the sight of her brain in that metal basin. That night sleep remained a stranger.

Back at the hospital the next morning, I took the stairs two at a time up to the ICU. Jennifer had the last room at the end of a long hallway. Every room but hers had bright artificial light emanating from it for the convenience of the nurses and doctors. From the far end of the hall, however, I saw no light pouring from her doorway. Her room was dim, which could have been a very good sign, meaning that she was awakened and was sensitive to light. Or it could have been bad.

I crossed the threshold of her room, ICU 8. A nurse, a resident, and Jennifer's parents turned to see who had entered. Behind them, on the bed, I saw Jennifer. Her breathing tube was gone, her eyes were open, and she was looking at me. I had never seen Jennifer awake before this. She had been fully sedated when she arrived with her parents a few days earlier. Only now did I see that her eyes were hazel-colored.

Sometimes an introduction isn't necessary. She could tell by the way her parents shook my hands that I was an indelible part of her journey.

I asked her two questions.

"Jennifer, can you please raise your arms up to the ceiling?"

She raised the right arm, but the left one lay flaccid next to her hip.

"Why can't I move it?" she asked me but with difficulty. Even the left side of her mouth was paralyzed. None of it made sense to her. She never had a chance to agree to a brain amputation because she had been so heavily sedated.

"Your parents asked me to help get rid of those horrible feelings you were having," I told her. "So I did a surgery that is going to make it hard for you to move your left side for a while. Now can you raise your legs for me?"

Again, only the right leg moved.

She started to cry. "When can I move them again?" she asked.

"I'm . . . not sure," I told her.

Her parents knew the one-sided paralysis was to be expected when they agreed to the hemispherectomy. Jennifer was never in on the decision. She never asked me to help her, and now I had to explain how I hurt her.

Her parents forced a smile to soothe her. They thanked me repeatedly: they hadn't seen Jennifer this alert and free of convulsions in months.

And then I went to my office, closed the door, and stewed. I was disgusted with myself. I paralyzed a little girl. I took half her beautiful brain and put it in a steel basin.

Welcome to my tortured Tuesday morning.

NEURO GEEK: THE HANDS
THAT WENT BLIND

A few months before I finished medical school rotations at Los Angeles County General Hospital in the spring of 2000, and before modern brain imaging was commonplace, I came across a wild paper in a journal. It was a time when the old orthodox view — that the function of every neuron is set at birth — was just giving way to our new understanding of neuronal plasticity.

The paper described the remarkable case of a sixty-three-year-old woman, blind since birth, who learned Braille at age six and later used it in college and at work. One day she told her coworkers that she felt light-headed and had difficulty; soon after, she passed out and was taken by ambulance to the hospital. After a couple days, she felt better, but when she tried to read a Braille "get well" card sent to her, she couldn't understand it. She said the Braille dots felt "flat" to her, or as if her fingers were covered by thick gloves. Yet she could still identify her house key by touch and could distinguish between pennies, dimes, and nickels.

What had caused her sudden inability to read Braille? An MRI of her brain showed that she had suffered a stroke in both of her occipital lobes — the section normally involved in vision. Yet, as noted in chapter 1, the sense of touch is usually controlled by the parietal lobes — the ones mapped by Penfield. Somehow, because she was blind, her brain had reassigned the vacant property of the occipital lobes to enable her fingertips to read Braille. All other uses of her fingers were still interpreted by her parietal; but for her, Braille was read with the same occipital region that sighted people use to read with their eyes.

The woman never regained her ability to read Braille. Instead, she used a computer with voice recognition software, and so was able to keep her job and remain productive.

HEALING

Over the next few weeks, Jennifer's incision began to heal, but she remained upset by the continuing paralysis, as did I. Psychologists and social workers visited with her, and I spent some time sitting and talking with her every day. I think we both knew I was doing it as much for me as for her.

Finally, after a few weeks, it was time for her to return home with her parents, a thousand miles away. Doctors and physical therapists there would take over the care. They didn't need us anymore.

Cards and emails kept me up to date on her recovery for a while. She stayed on an antiseizure medication, but at a low dose and without further seizures. She returned to school, one grade behind where she had been.

And then, three years after the surgery, an email arrived with an attached video file. I opened it on my phone and watched. One of her parents had recorded Jennifer, now nine years old, wearing a backpack and walking toward the camera. She was walking normally. It lasted just nine or so seconds. I could hear her laughing in it, though. I picked up that she still had a little droop on the left side of her mouth on the video. The accompanying email said she was doing well, even playing soccer.

So this time we had all been reprieved. Somehow, her brain had figured out a way to take over control of the left side of her body. With only half a brain, she remained a whole person.

NEURO GYM: BUILD YOUR NEUROPLASTICITY

Very early in my surgical training, I received some interesting advice from a senior surgeon. She told me surgery was a two-handed craft and suggested I spend my next vacation week with my right arm (I was right-handed) in a sling, even though it was uninjured.

I did just that, and I remember how relatively uncoordinated I was with my nondominant left hand versus my right hand. But I also remember how quickly my left hand adapted. Since then, I have actively tried to become ambidextrous in everyday life and still use a left-handed mouse, as well as use chopsticks and my smartphone with my left hand. This serves me well in the operating room, and anatomically, it keeps the movement cortices of both of my frontal lobes engaged.

But this functional plasticity is trivial when compared with the amazing recoveries I have seen my patients make. Some of them have left the hospital with loss of their right hand and weeks later returned facile with their left. I have seen others with legs that couldn't move immediately after surgery come walking in with a cane six months later. And I have cared for several patients with profoundly injured speech after surgery who return fluent after weeks or even months. Their stories are those of hard work and persistence in various neuro-rehabilitation clinics and facilities, facilities that exist because neuro-plasticity, with which existing neurons can assume new roles, is part of the brain's backup plan. I'm still amazed by the recovery potential of the injured brain and want to inspire you with the possibilities of what we can do to keep our brains fully engaged. Here are some ways to enhance your cognitive capacities and build your resilience by nurturing your brain's natural capacity for neuroplasticity:

1. **USE YOUR NONDOMINANT HAND.** If you aren't ambidextrous, try to do more with the nondominant hand. This will force the movement areas of your cerebral cortex to recruit idle neurons to the task. Learning to play a musical instrument is a great way to engage both hands individually and in concert.

2. **ACQUIRE A NEW LANGUAGE.** Learning a new language, even if you don't end up mastering it, is an excellent way to exercise the neuroplasticity of your left temporal lobe. As described in chapter 3, "The Seat of Language," our ability to communicate arises from the function of a vaguely delineated region; and the more square footage you acquire for it, the more cognitive reserve you'll have to access as you age.

3. **DON'T PRESS "ROUTE" ON YOUR PHONE'S MAP.** The main area for memory in the brain, the hippocampus, is also your brain's GPS. In fact, there are unique neurons (called grid cells) for navigating your way through a city or a subway. Tellingly, grid cells are part of the neuronal tissue that is lost in Alzheimer's, leading to disorientation in severe cases. So, using your internal compass rather than immediately pressing "route" on Google Maps is a great way to develop valuable spatial orientation skills.

11

THE BIONIC BRAIN

On a cool, sunny afternoon in the fall of 1964, Dr. José Manuel Rodriguez Delgado stepped from behind a wooden barrier and into a bullring in Cordova, Spain. Wearing a tie and sweater, Delgado carried neither sword nor cape. Standing alone on the dusty red earth beneath a deep blue Andalusian sky, he held only a small metal box that he had designed and built himself.

Released from its cage on the other side of the ring, a 600-pound bull named Lucero charged the physiologist. Its horns appeared to grow as it approached near enough that Delgado could smell the animal's musky scent. Holding his ground, Delgado waited until the bull was within eight feet before pressing a button on the metal box.

The bull skidded to a stop, its hind legs kicking up dust. So close that Delgado could almost touch him, Lucero blinked his eyes and breathed normally, as though a switch regulating his aggressiveness had been turned off in his brain.

In fact, it had.

A few days before, Delgado had sedated Lucero and another bull, Cayetano, by shooting them with a compressed-air tranquilizer gun.

With the assistance of the cowboys who worked on the ranch where the bullring was located, Delgado placed a small cage, known as a stereotaxic device, around Lucero's head. Using surgical tools, he cut through the bull's scalp, chiseled an inch-wide hole through the top of the bull's skull, and implanted a thin wire with two dozen electrodes at the end into Lucero's primary motor cortex. He repeated the process to place a second wire into the bull's caudate nucleus, and a third into his thalamus. Each wire was connected to a small radio receiver that was taped to one of the bull's horns. The hole in the skull was closed with dental cement, the skin surrounding the incision site was sutured shut, and the stereotaxic cage around Lucero's head was removed. Then Delgado repeated the process with the other bull.

A few days later, Delgado began his tests. With Cayetano standing peacefully by itself in the ring and Delgado safely behind the wooden barrier, he set his radio transmitter to activate only the electrodes attached to Cayetano's left caudate nucleus. At a very low 0.1 milliamp of stimulation, the bull showed no response. Delgado turned it up to 0.5 milliamp; Cayetano moved his head to the left. At 0.7 milliamp, the bull began to slowly walk to his left in a tight circle. At 0.9 milliamp, the bull repeated the circular motion — but faster. When Delgado changed the settings to stimulate the bull's *right* caudate nucleus, Cayetano spun again, but this time in a rightward direction. Throughout, the bull didn't moo, flee, or show any sign of agitation.

Next up was Lucero. This time, when the bull snorted and began its charge, Delgado sent 1 milliamp of stimulation to Lucero's caudate and thalamus, stopping Lucero in his tracks. Rather than just controlling movements, as with Cayetano, Delgado seemed to be pacifying Lucero as long as the button remained pressed.

After doing the experiment without the traditional cape of a matador, Delgado asked to borrow one from a famous bullfighter, El Cordobés, who was observing the experiment.

"My personal ability with the cape had been tested sometimes in the rural festivals of my youth," Delgado later wrote. "With the cape in my right hand and the radio transmitter in the left, I met with the bull Lucero, trying to keep my blood cool, although my heart beat with greater violence than I desired. I can tell you that one time there was a failure in the transmission circuit, and the bull managed to reach me, fortunately without more consequence than a good scare."

This was hardly Delgado's first experiment in the radio control of animals. For fifteen years, as a researcher at Yale University, he had been turning monkeys and cats into "little electronic toys." With the right placement of electrodes and the proper electrical stimulation, he could make them play, fight, mate, sleep, yawn, or bare their teeth. As quoted in a front-page article in the *New York Times* on May 17, 1965, Delgado said he had proved that "functions traditionally related to the psyche, such as friendliness, pleasure or verbal expression, can be induced, modified and inhibited by direct electrical stimulation of the brain."

The remarkable effects of electricity on the nervous system have been known since 1771, when Italian physicist Luigi Galvani discovered that an electric charge applied to a dead frog's leg would make it kick. He called it an example of "animal electricity," but others coined the word *galvanism*. Public demonstrations by Galvani's nephew, Giovanni Aldini, of the effects of electricity on dead animals and even on an executed prisoner at London's Newgate prison in 1803 (one eye opened, the right hand was "raised and clenched, and the legs and thighs were set in motion," according to a contemporary account) are believed by some to have inspired Mary Shelley's *Frankenstein*.

Disturbing as such experiments might sound, Delgado believed that his results could help humanity overcome its worst instincts. He eventually published his views in a book that, despite his best intentions, sounded to many like an Orwellian nightmare: *Physical Control of the Mind: Toward a Psychocivilized Society*.

Delgado was the very first person to implant a radio-controlled electronic stimulator into the brain, leading to the development of a procedure that has now been performed in more than 100,000 people around the world. He was at the forefront of a movement that has led to the implantation of all manner of electronic devices into the human brain to treat an astonishing array of diseases, rescuing ailing people with brain-machine interfaces.

Even so, it was not Delgado I was thinking of when a patient named Raymond came to my clinic a few years ago, asking if I could treat his peculiar affliction. I thought of Lucero, the bull. Ever since I heard of Delgado's twisted experiments back in medical school, it was always the bull I rooted for.

INVASION OF THE EYEBALL MICROBES

"I'm a banker," Raymond told me.

I had completed a neurological exam, and we now sat in an office. He was forty-five, Latino, with deeply furrowed creases around his eyes.

"I mean I *used* to be a banker," he continued, and pulled out a small plastic container of Visine eyedrops from the breast pocket of his suit jacket. "Pardon me," he said, and applied two drops to each of his eyes then returned the container to his pocket.

Raymond explained that he had always been mildly OCD: the sort of person who never left a piece of paper on his office desk, never a dirty dish in the sink. His wife, however, had been less meticulous — which is why he was in some ways relieved when she left him.

"That was two years ago," he said, and again pulled out the Visine, again applied drops to his eyes. "And that's when things got bad."

He had become obsessed with the belief that germs were entering his body through the mucosal membranes of his eyes. Hundreds of times a day, uncontrollable mental images had him compulsively using eyedrops. These intrusive thoughts were eating away at him and pushed him into depression.

"It makes no sense," he said, and pulled out the Visine again. "I know that. It'd be easier if I didn't know how crazy this is. But... I just can't stop."

A psychiatrist had prescribed all of the usual antidepressants and anti-anxiety medications known to relieve the symptoms of OCD: first Prozac, then Zoloft, Paxil, Lexapro, Celexa. None of them had worked, and he sought further help.

"I have nowhere else to turn," he said as Visine dripped down his cheek. "I'll try anything."

Having done research online, he wanted me to do something for him that was not so different from what Dr. Delgado had done to Lucero: to implant electrodes deep into his brain to send a signal that would pacify his OCD. To tame his demons, he needed mind surgery. A surgery not to cut out tumors or remove blood clots, but an operation that would alter the electrical oscillations flowing in his mind.

DEEPLY STIMULATING

First approved by the FDA in 1997 for the treatment of essential tremor and Parkinson's disease, deep brain stimulation (DBS) was eventually approved for OCD in 2009 — but only for severe cases in which medicines have failed to bring relief. It has also been used, although not yet with FDA approval, for treatment of chronic pain, major depression, and Tourette's syndrome.

How or why it works remains, for now, a mystery. Delgado didn't know why it worked on the bulls, and neither did Scribonius Largus, the physician of Roman emperor Claudius, who in 46 AD suggested that headaches could be treated by applying an electric ray fish to the scalp.

What we hypothesize is that the current discharged from the electrode somehow blocks or regulates the abnormal signals coming from nearby neurons. All we know for sure is that, in most cases, people suffering from a variety of neurological ailments do better when a stimulator is implanted and switched on.

All DBS systems have three parts: a battery-powered generator, which sends the electric signal; the electrodes, which deliver that signal to the brain; and an insulated wire to connect the two. Typically, the generator is surgically placed under the collar bone, and the wire runs (beneath the skin) up the neck, behind the ear, and down through the skull. The generator is then switched on and controlled with a hand-held wand.

Where exactly to place the electrode is a question that neurosurgeons continue to study. For the rigidity and tremor associated with Parkinson's, it is usually placed in one of two clusters of gray matter located well below the cortex and closely involved in the regulation of movement: the subthalamic nucleus, or the globus pallidus. For OCD, it can also be placed in the subthalamic nucleus (actually there are two of them, one on each side of the brain) because the two tiny lens-shaped clusters of neurons seem to be involved not just in controlling physical movement, but also in regulating *all* voluntary impulses — including, I hoped for Raymond's sake, the impulse to apply Visine to his eyes.

The operating room that morning included three nurses, a second neurosurgeon to assist me, the anesthesiologist, and an electrophysiologist. Just as Delgado did with his bulls, we began by affixing a stereotaxic cage around Raymond's head. The cage is necessary for two reasons: first, to form the outer grid lines for a precise 3-D map of his brain when he is placed in an MRI; and second, to keep his head perfectly still inside those grid lines during the surgery.

To attach the cage, four spots on Raymond's head were numbed up: two on his forehead above each eye, and two in the back of his head. Then I turned the four screws on the cage into each of those spots until they grabbed the outermost part of the skull. It looks undeniably barbaric, but patients tell me it's more intimidating than it is painful. We wheeled him to the MRI, took 3-D pictures of his brain inside the cage, and returned to the OR. I brought up the 3-D map on a computer and selected the right and left subthalamic nucleus as the targets. They are small and not dead center, so it's not an easy target. We selected the safest two approximately eight- to nine-centimeter routes to get to each one from the deep clusters of gray matter that govern your instincts and inclinations, and by avoiding, as best as possible, any sensitive brain areas along the way.

With route in hand, I knew exactly where to shave two small patches of hair on Raymond's scalp. I sliced C-shaped incisions and parted the scalp in each spot, drilled inch-wide circles in his skull, and then removed the silver-dollar-sized plugs of bone.

Now it was time to plan the trajectory, beginning with the left brain. I positioned a platform for inserting the electrodes directly over where I wanted the wire to enter Raymond's brain. The platform looks somewhat like a miniature derrick for drilling oil, and once in place, it plunged the electrode into Raymond's brain.

The spot it reached was as close as possible to the subthalamic nucleus, based on the MRI map, but close does not count in brain surgery. Especially with this brain surgery, where millimeters matter. We needed to hit the subthalamic nucleus exactly.

This is when the electrophysiologist took over as an ally to our mapping to confirm we were in the sweet spot. Using the electrodes as a kind of microphone, he listened to the noise generated by electrical activity of the nearby neurons, playing it loud on his computer. The rest of us stood still and listened along. To most it would sound like the crackle of static on a bad phone connection, but he had been trained to recognize the unique patterns generated by each cluster of neurons, just as birders can distinguish the chirp of a goldfinch from a white-eyed vireo.

"Advance one millimeter," he said to me.

I set the device to go one-tenth of a centimeter deeper into Raymond's brain. The pattern of static sounded unchanged.

"Another," he said.

Three more times we went steadily, ever so slightly deeper. And then even I could hear the pattern of static change. It was the brain's Morse code.

"Right there, bull's-eye," said the electrophysiologist.

About an hour later, after the other electrode had been placed at his right subthalamic nucleus, I removed the head frame and closed the incisions on his scalp with absorbable stitches.

Around 4 p.m., stopping by after my day was done to see how he was doing, I heard Raymond before seeing him. He was crying profusely.

"Raymond, you okay?!" I asked.

"Better," he said through the tears. "The anxiety — much less."

"But you're crying."

"Doc, I'm feeling better, actually pretty good, but the tears are like a leaky faucet I can't turn off."

Uncontrollable crying is known to be a rare complication of DBS. Other possible (though rare) side effects include hypersexuality, apathy, depression, hallucinations, a severe drop in IQ, and even euphoria.

We waited three days to see if this would change, but it did not. He no longer felt compelled to use Visine, but the crying wouldn't subside — and ironically in the extreme — he was now making eyedrops of his own.

We had to go back in. The whole procedure was repeated, from the MRI and the drilling derrick to the electrophysiologist listening for the unique neuronal chirps. In the end, we went a millimeter deeper.

When Raymond came to, the tears were gone along with much of his anxiety. A month later, when he came in for a follow-up exam, he brought no Visine. The whole experience reminded me that calling DBS a brain pacemaker is an oversimplification. It's actually mind control.

NEURO GEEK: MEMORY BOOSTER?

One of the most tantalizing treatments that DBS might provide is the enhancement of memory. For over a decade, DBS has been studied in healthy people as well as in those with early signs of Alzheimer's disease.

One of the first experiments to describe improvements in memory was originally designed to suppress appetite in an obese middle-aged man. As the study in *Annals of Neurology* described what happened when the stimulator was switched on: "The patient reported sudden sensations that he described as 'déjà vu' with stimulation of the first contact tested. He reported the sudden perception of being in a park with friends, a familiar scene to him. He felt he was younger, around 20 years old. He recognized his epoch-appropriate girlfriend among the people. He did not see himself in the scene but instead was an observer. The scene was in color, people were wearing identifiable clothes and were talking but he could not decipher what they were saying. As the stimulation intensity was increased from 3.0 V to 5.0 V, he reported that the details in the scene became more vivid. The same perceptions were obtained during blinded, sequential successive stimulation of individual contacts."

The electrodes had been placed near the fornix, where nerve fibers cross paths with the region in our brains where memory is formed: the hippocampus. With the voltage turned down halfway, the stimulation continued to result in major improvements in verbal memory for months afterward. (It did not, however, help the man lose weight.)

Following up on that study, a team of researchers at UCLA and Tel Aviv University in Israel tested DBS of the hippocampus and surrounding regions in seven patients. They found that it improved performance in a memory task by a remarkable 64 percent. But a larger study involving forty-nine people found the opposite effect: stimulation actually impaired memory. Technical differences between the two studies, however, left researchers unsure what to make of the conflicting results.

Then came a report in 2018 by a team led by Michael J. Kahana, director of the University of Pennsylvania's Computational Memory Lab. Rather than stimulating the hippocampus, his group stimulated another area where memory is encoded: the left temporal cortex. Instead of simply zapping the brain with the usual stimulation used in DBS, his group first recorded electrical activity in the area while people were taking memory tests.

They found two unique sets of electrical patterns: a "smart" one, when people were doing slightly better on their memory tests, and a "dumb" one, when they were doing slightly worse. Then, using the electrodes to play back the "smart" pattern while the people took another test, they consistently found a 15 percent improvement in recall.

That might not sound like a lot, but it is about equal to the *loss* of memory that occurs in a person with Alzheimer's disease over the course of two and a half years.

"Taken together," Kahana's paper concluded, "our results suggest that such systems may provide a therapeutic approach for treating memory dysfunction."

Brain surgery for the 5.7 million Americans with Alzheimer's disease is not in the cards, however, given that fewer than a million brain surgeries of *any* kind are performed each year in the United States. That's why treatments using external electrical stimulation without surgery sound so appealing.

OVERCOMING PARALYSIS

Other types of brain implants are being used to restore movement and sensation in people paralyzed due to spinal cord injuries. Called neuro-prosthetics, these devices work to go around the spinal injury by detecting the brain's signals for a particular movement and then transmitting those signals directly to the nerves controlling the arms, legs, or hands.

BrainGate, one of the most promising devices, drew headlines recently for its extraordinary success with a fifty-three-year-old man who was paralyzed from the shoulders down. Eight years after a bicycle accident left him unable to use his hands or legs, two electrodes were implanted in his motor cortex and another thirty-six electrodes in his arm. With the BrainGate system translating his signals from his brain and transmitting them to his arm, he trained for a year, eventually learning how to reach and grasp with enough dexterity to eat with a utensil, drink from a cup, or scratch an itch.

Because all our movements are refined by the feelings we get from our legs, hands, fingers, and body as we move, restoring sensation is hugely important if we are ever going to restore natural movement. Incredibly, that bridge has been crossed, too. Researchers from the University of Chicago, Case Western, the University of Pennsylvania, and elsewhere built two microelectrode arrays, each one smaller than a pencil eraser yet each bristling with thirty-two tiny electrodes. Working with a twenty-eight-year-old volunteer named Nathan Copeland, who had quadriplegia following a car accident in 2004, they placed the electrode arrays on his parietal lobe, where feelings of the index and little fingers are processed by the brain. They then linked those electrodes to sensors in a prosthetic limb.

At first, Mr. Copeland felt spontaneous tingling in the prosthetic limb's fingers. But after a few weeks, he felt them only when the limb's fingers were touched or pushed.

"I can feel just about every finger — it's a really weird sensation," he said. "Sometimes it feels electrical and sometimes it's pressure, but for the most part, I can tell most of the fingers with definite precision. It feels like my fingers are getting touched or pushed."

A two-way brain-machine interface may sound like science fiction, but it's simply the next step toward our bionic future.

NEURO GYM: RESET YOUR VAGAL TONE

There are twelve unique nerves that sprout from the brain, exit the skull into the face, and mediate smell, sight, hearing, taste. These are called cranial nerves, and the tenth one is nicknamed the "wandering nerve" because it's the only one out of the twelve that ventures well beyond your face and around your heart and lungs. After its exit from a tiny hole in the base of the skull, it descends in the neck tucked between the carotid artery and jugular vein. If you were to slice it and look at its cut end head-on, you'd see fibers called efferents, because they carry outbound signals to your chest cavity. These are half of the nerve fibers that your brain uses to control your heart rate in times of stress or rest; specifically, these are the fibers which are activated in rest. They are also, unsurprisingly, the pathway that Buddhist monks and others, through years of training, use to lower their heart rate with thought alone.

Less well known is that the wandering nerve is a two-way street. It also carries signals (via afferent fibers) back into the skull from your heart and lungs to shower the brain with ascending information signaling the brain to enter a more calm and restful state. And these fibers can be manipulated or even hijacked. How? You can do it yourself through the intense practice of mindful breathing. Meditative breathing can calm the electrical oscillations and stress responses in your mind by resetting your vagal tone. The neuroscience term is "network reset." So, despite the increasing popularity of inserting catheters and wires into brains, the basic ability to modify your thoughts and feelings is actually already a built-in technology in each of us. I'm not saying that mindful, meditative breathing alone would have necessarily relieved Raymond's compulsive use of eye-drops; certainly, it would not have helped Nathan Copeland regain feeling in his fingers. But I do say this: never underestimate the power of the human brain.

NEURO BUSTED: MENTAL TELEPATHY

Elon Musk spent $27 million of his own money to start a company called Neuralink. On its bare-bones website, the company states: "Neuralink is developing ultra-high bandwidth brain-machine interfaces to connect humans and computers." At a speech he gave in Dubai, Musk said: "Over time I think we will probably see a closer merger of biological intelligence and digital intelligence. It's mostly about the bandwidth, the speed of the connection between your brain and the digital version of yourself, particularly output."

Blogger Tim Urban spoke with Musk about the company's goals for building a "wizard hat." How long might it take for such a breakthrough to occur? "I think we are about eight to ten years away from this being usable by people with no disability," Musk said. But to date, the only direct brain-to-brain communication ever described in the scientific literature—the closest to mental telepathy yet achieved—was incredibly convoluted and painfully slow. Developed by researchers from Harvard and a Spanish company called Starlab, using a combination of EEG and transcranial magnetic stimulation, the communication took hours to transmit a four-letter word. It's highly unlikely mental telepathy is in our future, and many wonder if this technology is even needed.

12

SHOCK AND TINGLE

According to the top line of her chart, Mrs. Chang was a sixty-eight-year-old woman with a forty-year history of bipolar disorder with catatonia. I stood outside the room, reading the chart and a letter from her referring psychiatrist.

It was 1999, my third year of medical school, and I was moving through the required three-month rotations, of which psychiatry is one. This was the year we finally got a chance to actually engage patients.

Psych was a strange one. Riveting stories, tortured souls, but not much to do for them was the reputation among students. I saw schizophrenics who refused to take their medicines, teenagers starving themselves due to anorexia, convicts who feigned illness to get out of the penitentiary for a few nights — the breadth of humanity. Until I had started my three-month psych rotation, I knew of these diseases only from books, but now I was expected to help out residents and attendings who had chosen to make psychiatry their specialty. This morning, my job was to find out what was going on with Mrs. Chang.

I entered the room to find three adults seated in chairs: a man and a woman in their midforties and the woman who had to be Mrs. Chang. She was stooped, wearing a jacket, with hair jet-black except for a few electric strands of white above her temples. She gave no notice to me, staring straight ahead. The younger man and woman with her stood, introducing themselves as Mrs. Chang's son and daughter. The three of us sat down, and I turned to their mother.

"How are you feeling?"

Her eyes remained fixed on the wall across from her. Her gaze never broke.

I turned to the daughter. "I saw the letter from your mother's psychiatrist," I said. "Please share more with me about her and how things have gotten so difficult."

Their mother had first been diagnosed in Shanghai, she said. As children, they would see her lying in bed sometimes for days or weeks. Other times she became hyper and started giving away money to strangers. Sometimes she wouldn't sleep for days, brimming with energy. She had initially been given traditional Chinese medicines, and then their late father, a sophisticated man, had managed to get her a prescription for lithium in 1968, just as it was beginning to catch on as a treatment around the world. When the family moved to the United States in the late 1970s, she continued on lithium, the standard treatment for bipolar disorder, and then added Prozac when it became available in 1988. That had mostly worked, until now.

"She's never been this bad," the son said.

"It's like she's dead," the daughter said.

"So we want to try shock therapy," the son said. "Her doctor recommended it. That's why we're here."

Through our conversation, Mrs. Chang was like a stone. I barely took notes because I wanted to hold her children's eyes and try to read their emotions.

"It's like there is cancer in her mind," the daughter said. "And now she can't even feed herself, and she refuses to let us feed her. She's lost seventeen pounds. We don't know what to do."

The son reigned me in with his fatigued but resolute eyes, and said, "If it was your mother, would you let them shock her?"

I knew that electroconvulsive therapy (ECT), the technical term for shock therapy, was still being used at that time, but I thought of it as a last-ditch relic from a bygone era. Like most people back then, I associated it with actor Jack Nicholson's devastating performance in the 1971 movie *One Flew Over the Cuckoo's Nest,* in which he is forced to have the treatment at a psychiatric hospital as a kind of punishment, held down by orderlies as he violently convulsed. It had the patina of cruelty in my mind.

Before I had a chance to answer the son's question, however, the door opened, and the attending psychiatrist walked in.

"Rahul," she said, "give me a brief rundown on Mrs. Chang."

NEURO BUSTED: THINKING CAPS

Another form of brain stimulation involving so little electricity that a person can barely feel it is fast gaining support from researchers around the world. Transcranial direct-current stimulation (tDCS) involves the placement of an electrode-studded skullcap over a person's head and the discharge of an extremely low dose of electricity. The treatment is given for perhaps ten minutes, but the effects have been seen to last for weeks or longer. Many studies have shown that it might relieve depression, improve memory, enhance creativity, and boost attention in sleep-deprived people far better than caffeine does.

But the emphasis should be on the word *might*. Not all studies have found benefits. One of the largest and most recent, from an international group including researchers at the National Institutes of

Health, involved 130 people with depression. Ridiculously, more people actually felt better after receiving the placebo version of tDCS — with a level of electricity so low as to be considered useless — than after getting the real thing.

Its effects on memory, creativity, and intellectual firepower look far more consistent. During a single treatment or in the weeks following a series of treatments, tDCS has been shown time and again to improve memory and problem-solving abilities. Not every single study has found a benefit, I should point out, but most have found measurable improvements whether the people being tested are healthy, high-functioning young adults or older people with a neurological disease that impedes memory.

A researcher at Oxford University in Britain has even found that a type of tDCS improved the math skills of college students — six *months* after they had undergone stimulation.

What I find concerning about this field of research is that tDCS devices are being bought online and used as a DIY treatment by self-professed "brain hackers."

Pardon the pun, but anyone who tries stimulating their brain at home needs to have his or her head examined. These devices require placement of the stimulators to precise areas of the skull with precise dosing. Using it for longer than recommended could have bad effects. As with any treatment, too much can harm.

Is the research exciting? You bet. Does it look promising? Very. Should you try it at home, on your own, without medical supervision? I wouldn't.

LIFE-SAVING

Many people still don't understand what I have since learned: that ECT, when properly administered, is the fastest, most effective treatment for depression and bipolar disorder when medications have failed to work. Up to 90 percent of people with major depression for whom nothing else worked find relief in days or weeks, and the effects on memory are usually (but not always) modest. It is also sometimes given soon after the onset of schizophrenia. And unlike in *Cuckoo's Nest,* people no longer convulse during the treatment because they are given medicine beforehand that temporarily paralyzes their muscles, much like being under anesthesia for surgery.

But nobody should deny that ECT's bad reputation of fifty years ago was well deserved. The doses of electricity given back then were so much higher that people were often utterly disoriented the following day, unable to remember even their names. And, long-term, some people lost their memory of the prior year or more.

With doses now significantly scaled back, most people have only mild, temporary memory loss. Exceptions do occur, but that risk must be balanced against the 20 percent risk of suicide in people who remain severely depressed. And new treatments, such as magnetic therapy, are being studied as a way to relieve depression without almost any effect on memory at all, by avoiding stimulation of the hippocampus. The main catch is that relief lasts only about six months, so some people need follow-up treatments.

Why it works remains unclear. Some believe it kicks abnormal electrical oscillations across the brain — the kind that can be measured with EEG — into a healthier, more normal rhythm. But that's still just a theory. What's clear is that it *does* work when medicines fail. Yet to this day, only about 10 percent of the people who could benefit ever get it.

Why so few? Primarily because of its lingering negative reputation even among doctors. After all, how strange is it that seizures, usually treated as a disease, can be used as a kind of medicine to treat another disease, depression? For a neurosurgeon, aberrant electricity is something to extinguish. For psychiatrists, that same pulse of electricity can be a therapy, a tool that can jolt neurons into a mind reset.

NEURO GEEK: ELECTRICITY AS MEDICINE

Almost as soon as experimenters learned how to generate an electrical charge in the 1700s, they began to investigate its therapeutic uses. Benjamin Franklin claimed he cured a woman of "hysterical fits" by placing her near a machine that generated static electricity. In 1744, the first issue of *Electricity and Medicine* was published. But these were mild charges, far less than necessary to induce convulsions.

It was insulin, of all things, that was first used to put mentally ill people into convulsions in an effort to cure them. Beginning just six years after its discovery in 1921, insulin was given in large doses to people with schizophrenia by psychiatrist Manfred Sakel. He claimed that 88 percent of his patients improved dramatically after receiving enough insulin to cause temporary convulsions due to an extremely low blood sugar level. Soon after, another drug, metrazol, was also found to induce convulsions. Not until 1938 did an Italian psychiatrist, Ugo Cerletti, describe the use of a strong electrical charge to the brain as a new, faster way to get the same effect. Cerletti's claims were greeted in the United States with skepticism until a Cincinnati psychiatrist gave a public demonstration of ECT in 1940 during a meeting of the American Psychiatric Association.

Soon after, it became the standard, "modern" treatment for major depression.

Amazing, isn't it, how a totally bogus therapy — Ben Franklin's electrical cure for the "fits" — turned out to be a highly effective treatment for depression and bipolar disease when given at doses just high enough to induce a convulsion? You can't make this stuff up.

ANTICLIMAX

Two weeks after seeing Mrs. Chang and speaking with her children, I joined the chief psychiatric resident, an anesthesiologist, and a nurse in the ECT procedure. Still catatonic, seemingly insensate to the world around her, Mrs. Chang was wheeled in on a gurney with an IV dangling from a vein in her left arm.

White circular stickers, each housing a tiny metal pellet, were attached to her forehead, with ointment to dissipate heat. One was placed on each temple to deliver the electricity. Two more were placed to read the ensuing brain rhythms. When fully prepped, her face was festooned with wires.

My job as a student was making sure she didn't bite her tongue. Even though the anesthesiologist was giving her succinylcholine, a muscle paralytic, the jaws clench forcefully at the onset, one last throe of activity before becoming dormant. The attending directed me to take a stack of square cotton gauze pads, we call them 4 x 4s, roll them into a loose cigar, and slide them into her mouth on one side.

The ECT machine was wheeled over. It looked remarkably unimpressive: about the size of a toaster oven with buttons and a dial. The matte gun-metal finish had no frills.

The attending set the dials to deliver half a millisecond of 550 milliamps. (That's about enough electricity to light a 60-watt bulb, but for only one-half of a thousandth of a second.)

Then, with a nod to us, the attending pressed the button that delivered the electricity. Nothing dramatic was perceptible but for the occasional flutter of her toes and the twitch in her jaw. The heart monitors were steady, but the brain waves were now furious for a long moment. That was it.

We stood around until Mrs. Chang slowly woke up about fifteen minutes later. Her eyes remained fixed; she didn't speak. There was no apparent change in her condition.

I helped wheel her into a recovery room, where her son and daughter came to see her. She would get another treatment in two days, on Wednesday; a third treatment on Friday; and a few more treatments the following week. We would just have to wait and see how she did.

REBIRTH

Two weeks after Mrs. Chang had completed her six sessions of ECT treatment, I walked into the same exam room where I had first met her and her adult children. I had not seen her since that first session, when she had seemed so unchanged, so I had low expectations.

They sat in the same order, son on the right, daughter on the left. I shook the son's hand and went to shake the daughter's, but Mrs. Chang looked up at me and raised her hand. She caught me off guard, and then we had our first handshake.

It was as if shackles had been lifted from her mind. She went on to share wonderful moments from her life that were unrecoverable in the grip of bipolar disorder. Her verbal fluency surprised me; it was like when people have their speech function return after a partial injury to the left temporal lobe. But her brain had never been structurally injured, nor had it been repaired in that sense. Rather, the frayed electrical rhythms of her brain had simply been reset back to mental health.

I still am in awe of what I witnessed. The dramatic impact I now see with surgery was achieved without cutting. The effects of this once-vilified treatment are, like all great things in medicine and science, magic. A woman who had been unable to talk or move freely for months and who had spent decades suffering with bipolar disorder was freed from her mind's bind.

Now I knew how to answer the question that Mrs. Chang's son had put to me the first day we had met: "If it was your mother, would you let them shock her?"

I would. In a heartbeat.

NEURO GYM: NUTRITIONAL PSYCHIATRY

As I hope I've made clear, the ECT described in this chapter is only for people with major, disabling depression that has not responded to medications or talk therapy. For the far larger number of people with less severe forms of depression, it may be possible to influence your mood with the daily choices you make about nutrition. In chapter 9, I shared with you the power of food in the form of an all-fat ketogenic diet to break seizures in children. For them, food is medicine. Depression and anxiety are increasingly being understood in terms of aberrant electrical brain waves, and doctors in the emerging field of "nutritional psychiatry" are prescribing diets that can lessen anxiety and improve your mood.

What I find interesting and fortuitous is that the same kind of healthy diet being prescribed for better mental health overlaps almost exactly with the MIND diet, which staves off dementia. For instance, one study by researchers at Rush University Medical Center in Chicago found that older adults who eat fruits, vegetables, and whole grains are less likely than those who don't to develop depression. Even more impressively, two clinical trials have found that when depressed people are counseled by nutritionists on how to eat a healthier diet, their depression improves.

What kind of diet did the nutritionists recommend? You already know the answer: more fruits, vegetables, seafood, and whole grains; less meat, fried foods, carbs, and fat.

Ultimately, the lesson is that the "comfort foods" that many of us reach for when feeling blue — the meatloaf and mashed potatoes; the burger and fries — may not bring comfort to those most in need of it.

13

STEM CELLS AND BEYOND

I wasn't the first person to inject living stem cells into the brain of a person in hopes of killing cancer. In fact, I wasn't even the second. But I was the third.

As part of a first-in-human clinical trial led by my senior colleagues, neurosurgeons Dr. Behnam Badie and Dr. Mike Chen at City of Hope, we were responsible for not only performing the surgery, but also for injecting neural stem cells directly into the brain to chase down and eradicate escapee cancer cells.

Stem cells, as you probably know, are capable of turning into any mature cell. They are the progenitors of all the other cells in your body, whether bone, skin, brain, or muscle. Neural stem cells (NSCs) are just slightly more differentiated: they can become any type of cell found in the nervous system, including any variety of neuron in the brain.

One interesting quality of NSCs is that they naturally gravitate toward any brain tumor. What if, our team wondered, they could be tweaked to deliver chemotherapy selectively to the cancer while sparing the rest of the brain? For our study, the NSCs had their DNA manipulated to express an enzyme called cytosine deaminase.

Stay with me here.

The thing is, cytosine deaminase itself has no effect on cancer cells, but it *does* have the capacity to chemically switch an antifungal medicine, called 5-fluorocytosine, into a chemotherapy drug called 5-fluorouracil. This turns out to be useful in brain cancer because 5-fluorouracil cannot cross the blood-brain barrier, but 5-fluorocytosine can. So now we have an antifungal drug with relatively modest side effects, and it can be transformed into a chemotherapy drug only upon reaching the NSCs that clustered around the patients' brain tumors. Outsmarting cancer demands a creative approach.

Most of the fifteen patients who were part of the trial had been previously treated for glioblastoma, a rare but deadly brain tumor. How deadly? Only about 2 percent of patients with glioblastoma survive beyond two years even after surgery, chemotherapy, and radiation. Glioblastoma is the type of cancer that killed Massachusetts senator Ted Kennedy in 2009; Vice President Joe Biden's adult son, Beau Biden, in 2015; and Arizona senator John McCain in 2018. Because the tumors send out tentacles into the nooks and crannies of the nearby brain tissue, it's impossible for us to surgically remove it all.

Even so, we began the study with some optimism because in mice with glioblastoma, the treatment had cured them. But the experiment had not yet been done in humans. No promises of a "cure," however, were made to the patients. They were hoping to live longer as a result of our trial, but they also wanted to contribute to the advancement of science and medicine for others afflicted with this terrible disease. If it helped them survive even a few additional months, that would be a therapeutic advance not seen in decades.

The operations to open their skulls and implant 2 million modified neural stem cells took about four hours. Placed in a specialized test tube, the solution of cells looked mostly clear with whitish swirls. Once the tumor was dissected out, the needle attached to the syringe was inserted about a centimeter into the resection cavity and cells injected. Repeatedly the needle punctured and injected cells in ten to twelve different spots, into the crater left behind in the brain.

The excitement around the study was palpable. A *CBS Evening News* crew showed up on the first day that the stem cells were given to the first patient. We all knew that this first-in-humans study was historic. Being a part of something like this was the reason I chose City of Hope over Harvard and Stanford. At City of Hope, the forward-thinking environment didn't motivate us just to discover but to do so in the shortest time feasible.

The only question now was: Would the stem cells work in our patients anywhere near as well as they had in mice?

SLOW BUT STEADY PROGRESS

First of all, none of the patients experienced any significant side effects from the treatment, which is the first fundamental step in testing a new therapy. Before we determine whether or not it works, we have to show it's safe. And we did. If you think about placing genetically tweaked stem cells from one human into another, having nothing go wrong is in itself remarkable. Since cell therapy is in its infancy, this was an important result.

Second, the modified NSCs given to each volunteer did, as hoped, navigate their way to the remaining tumor cells. Third, upon their arrival, the NSCs succeeded in transforming the antifungal medicine given to the patients (5-fluorocytosine) into the chemotherapy drug 5-fluorouracil. Check, check, check.

But the question of greatest concern to the patients involved in the study, to their families, and to all of us involved in it was whether it would extend their lives. In a study of just fifteen people, such answers are hard to come by, but the findings were provocative. Those who received the low or middle doses lived, on average, just under three months after the procedure, while those given the highest drug dose lived, on average, for about fifteen and a half months. Another way of looking at it: only two of the nine patients on the lower doses lived longer than twelve months, compared to four of the six patients on the highest dose.

Still, to our great disappointment, all of the patients did die. You might say, therefore, that the study was a failure. But as someone who treats these kinds of cancers routinely and knows how grim the prognosis is for most of my patients, I considered it an important step forward. Just the fact that the stem cells didn't produce any bad side effects was a big deal. That they worked as predicted was, to me, impressive. And that they showed a hint of prolonging survival was amazing. We were treating a disease that has become the Mount Everest of challenges for oncologists. Every step upward gets us closer to our goal of better treatments for our patients. And the lessons learned along the way are relevant to the field of oncology as a whole.

I'm often asked if cancer can be cured. Well, it depends on which type of cancer we are talking about, since there are more than 200 types, and each can be diagnosed at various stages. Frankly, for the most aggressive cancers, such as glioblastoma, extending life by turning cancer into a chronic disease would be a huge victory. To achieve something similar to what medical scientists have to this point achieved with HIV — for which we still don't have a cure, but HIV patients are now living into old age — would be a major advance and seems a reasonable goal for this generation of cancer researchers. However, for other cancers, a cure is realistic. I'll get into this next.

A LIVING DRUG: IMMUNE THERAPY

Neural stem cells are not the only type of living cell therapy being used to treat brain cancer. I'm part of the team developing a kind of anti-tumor immune therapy using what's called chimeric antigen receptor (CAR)–engineered T-cells.

Let's break that down.

The word *chimeric* comes from Greek mythology, in which a chimera was a fire-breathing monster with the head of a lion, the body of a goat, and the tail of a snake. CAR T-cells are called "chimeric" because they begin with ordinary T-cells taken from a person's blood (the so-called "killer" white blood cells that recognize and attack a broad range of bacteria or viruses) and attach a kind of lion's head: a molecular signal that guides the T-cells to seek and destroy cancer. The T-cell part of the chimera is a powerful killer, and the antibody at the head makes the attack more specific, reducing side effects.

To create the CAR T-cell therapy, we begin with T-cells taken from a patient's own blood. Next we add the patient's specific cancer receptors, making millions of copies of sensitized T-cells. Then these custom-built, supercharged immune cells are administered back into the patient's blood.

Research teams around the world have had incredible, unprecedented success in using CAR T-cell therapy on leukemia and other cancers of the blood. People thought to be on death's door have recovered and gone on to live for years with no sign of remaining cancer.

By the time this book is published, I expect that the CAR T-cell study will have already begun, with myself as one of the three surgeons. The brain cancer we will be treating is not glioblastoma, as mentioned above, but a different kind that is much more common: brain metastasis from breast cancer. These brain cancers originate from breast cancer cells that have escaped into the patient's blood vessels and then breach the brain's blood-brain barrier to take root as a new tumor.

Even with existing therapies, women with breast cancer that has metastasized to the brain typically live only a few years. In an effort to beat those odds, I will surgically insert tiny plastic tubes into the brain ventricles of women who volunteer to participate in the study, so that the CAR T-cells can be dripped directly into their cerebrospinal fluid. The little plastic tubes will remain in place for months to continuously enrich the brain's fluid lakes with CAR T-cell therapy.

As another first-in-human study, no one can know the outcome until the study is completed. But I am hopeful. If it works as well in this kind of brain cancer as it has in leukemia, the results will be game-changing for my patients battling advanced cancer.

NEURO GEEK: DOES THE ADULT BRAIN GROW NEW NEURONS?

It was in La Jolla, where in 2006 I did the experiments for my PhD, that a cornerstone of neuroscience orthodoxy was overturned just eight years earlier. In 1998, Fred "Rusty" Gage published a revolutionary paper in the journal *Nature Medicine*, showing that the hippocampus, where new memories form, generates new neurons on a daily basis from a small population of neural stem cells (NSCs). Until then, it had been believed for nearly a century that no new neurons are ever generated in the adult human brain.

This was first shown in birds. Birds had brain stem cells that they used to sprout new brain tissue and connections to flirt better and more competitively serenade each mating season. Even nonhuman primates, studies found, continue to grow new brain cells well into adulthood. But nobody had ever found them in older humans until Gage. Soon after, other teams found that new neurons are also generated from NSCs that line a wall of the ventricular plumbing inside the brain. The result has been twenty years of research aimed at figuring out how to prompt these NSCs to speed up their work, perhaps to replace neurons lost during Alzheimer's or Parkinson's disease, or due to cancer or traumatic injury.

Some studies found that exercise could increase the production of new neurons. Others used drugs, electrical stimulation, brain training—all kinds of things.

Then, in March 2018, news broke of a study out of UC San Francisco that claimed to find, after an exhaustive search, no signs of new brain cells after all.

Led by Arturo Alvarez-Buylla, a team of nineteen researchers that included scientists from Los Angeles, China, and Spain examined fifty-nine brains from fetal stages to age seventy-seven. Looking only at the hippocampus and expecting to find the estimated 1,400 new neurons that previous studies had said are born each day there, they instead found zero, despite using a variety of techniques designed to identify new neurons.

Gage immediately pushed back, noting that the people whose brains he had studied had agreed to take an imaging molecule that attaches only to newly divided cells as part of cancer treatment they were undergoing; upon their deaths, they found the molecule attached to hundreds of neurons.

Most of us took the paper from Alvarez-Buylla with a large grain of salt. Too many studies had been published finding neurogenesis in adults to be convinced by a single study that it was all an illusion. If my skepticism surprises you, remember that science is not religion: it moves forward only when doubting researchers (like me!) put assumptions and claims to the test. If Alvarez-Buylla's conclusions are replicated and all the earlier studies of neurogenesis turn out to have been wrong, I'll be the first to accept it. But until then, I'm keeping an open mind.

This debate is not over. For now I remain on the side of the majority of scientists who believe that the human brain keeps producing neurons, in select regions of the brain, throughout life.

One thing I am certain of, however, is that cell therapy of some kind — it might involve stem cells or CAR T — will eventually overcome cancer and other diseases of the brain. It might not be tomorrow. It might not be in my lifetime. But scientists will keep investigating. We will keep making progress. We are relentless.

HOPE FOR ALZHEIMER'S

Cancer is not the only brain disease being targeted for treatment with living cells. Cell therapy also has the potential to replace aging or injured tissue. In this hope for regenerative medicine, modified stem cells are being studied as a treatment for Alzheimer's disease.

Scientists were recently blown away by a study using many of the newest tools of biotechnology to attack the familial form of Alzheimer's in which people are destined to develop the disease at an early age, usually between fifty and sixty-five.

The study began with some skin cells taken from two sisters, both of whom carried a mutation in a gene, called presenilin 2, known to cause early Alzheimer's disease. Like their mother before them, both sisters were showing early signs of cognitive decline. The skin cells were put through a series of test-tube treatments that reverted them back into a type of stem cell called induced pluripotent stem cells, or iPSc. (Think of it as the Benjamin Button technique, named after the movie in which the title character, played by Brad Pitt, starts life as an old man and slowly "grows" into an infant.) Then they put those stem cells through another batch of treatments that nudged them to develop into a specific type of brain cell (called basal forebrain cholinergic neurons). It's a sort of cellular alchemy.

Finally, they corrected the mutation in the matured cells using the latest trick for editing genes, called CRISPR/Cas9. It works like a molecular scissor and stitch, permitting scientists to remove unwanted gene sequences and insert new ones, almost at will, like editing a Word document. First described in 2012 and applied to human cells for the first time in 2014, the technique is so new and transformational that it is hoped that all manner of mutations might one day be repaired.

So, in the study of the two sisters, researchers wanted to see how these corrected cells worked — *if* they worked, that is, and if they were safe. Rather than placing them directly into the sisters' brains, they placed the neurons into test tubes and watched how they behaved. To their delight, the corrected brain cells showed normal, healthy electrical properties and normal production of beta amyloid—the protein that, in people with Alzheimer's, gunks up the brain and prevents normal cell-to-cell communication.

The next step will be to test the treatment in mice, said the study's leader, Dr. Sam Gandy, of Mount Sinai School of Medicine. It probably won't be fast enough to help the two sisters who donated their skin cells for Dr. Gandy's study, but slow and steady is how science works.

NEURO BUSTED: BUYER BEWARE

The hype over stem cells has reached the point where you can find websites from hucksters offering to treat just about any disease with them. What's really frightening is that some of the companies offering this stuff have legitimate physicians serving on their medical boards and give every impression of knowing what they're doing.

I have seen these outfits with my own eyes. A few years ago, a foreign company that offers stem-cell treatments as a cure-all for just about anything asked me to visit their clinic in Kazakhstan. They were offering intravenous stem cell injections to anyone who was in search of relief from a wide range of conditions from arthritis to fatigue. Some were there with no medical conditions and in hopes of "rejuvenation." I guess they hoped I would give them some kind of endorsement, but I made them no promises.

Astana, the capital, is a planned city built nearly from scratch beginning in the final years of the twentieth century. Variously called a "space station in the steppes" or "the world's weirdest capital city," Astana is located in the middle of the largest and emptiest expanse of grassland in the world. Futuristic spires and skyscrapers make for a kitschy Tomorrowland vibe, yet the malls sell not only Cartier and

Prada but fermented horse milk and wooden clubs that could be used in *Game of Thrones*.

Upon my arrival, I was driven to the stem-cell clinic, a sprawling one-story building. Inside, it was clean and neat but strange, decorated with flowered curtains, doilies, and furniture from the 1970s.

The director of the clinic ushered me into a treatment room, where a white-haired gentleman sat with an IV in his arm. Handsome and healthy-looking, he told us that he had recently retired from his job as CEO of a major corporation. He wasn't sick, he insisted; he just wanted to put some pep in his step.

This wasn't science. This was crazy. Stem cells are not vitamins. They don't just magically make people feel better. How could someone smart enough to run a large corporation be so gullible?

Five other people were getting treated in other rooms. Two of them, like the CEO, came from the United States, one came from France, and another from China. Each had paid the equivalent of $20,000.

The director took me into a funky boardroom. I asked him where the stem cells come from, whether they follow the kind of manufacturing processes required in the United States, how they follow up on the people's outcomes. His answers might have sounded convincing to someone who doesn't know the science, but they made absolutely no sense to me.

I remained pleasant, polite, and professional. I didn't throw any chairs or scream at the people being treated to get out and run for their lives.

"This makes no scientific sense to me," was about all I told the clinic director.

But it really bothered me. Perhaps nobody was getting sick from the modern snake oil this company was pumping into people's arms; but the real danger, for people seeking medical treatment for an illness, is that they might get this instead of something that could actually help them. Taking people's money and giving them only false hope in return is a kind of injury.

It was one of the great honors of my career to participate in a pioneering stem-cell study. But there is still much that science does not know about when and how they can work to treat cancer and other diseases.

Please do not be manipulated by these standalone clinics that seek to profit off people's desperation. The difference between them and legitimate university-based medical institutions that actually know what they're doing is big enough to drive a hearse through.

MAKING THE BLIND SEE

We already know, from many of the studies above, that neural stem cells implanted into the brain can survive. We also know, at least in a test tube, that when they mature into functioning neurons, they appear to work normally. But a major remaining question is whether the corrected neurons will sprout the axons and dendrites necessary for them to communicate with other, existing neurons.

The answer to that question, based on a recent study, is a tentative yes.

You have heard of the three blind mice, chased by a homicidal farmer's wife wielding a carving knife?

Neurobiologist Andrew Huberman made blind mice see.

To begin his experiment, he crushed a type of neuron at the back of the eye called the retinal ganglion cell. These cells sit right up against the retina, where they take the light received by the eye and translate it into electrical nerve signals. They then send those signals down their axonal tail, all the way into the spot in the brain where the signals are processed so that they can be perceived as "vision" by the mouse.

Once the neuron was crushed, however, the mice became blind.

But not for long.

After crushing their retinal ganglion cells, Dr. Huberman used gene therapy to increase the nearby amount of a protein, called mTOR, which prompts cell growth and survival. At the same time, even though the mice were apparently blind, he exposed them to high-contrast visual stimulation for three weeks, using the image of a moving black-and-white grid.

With the two treatments combined — the gene therapy plus the visual stimulation — the mice began reacting normally to visual images; for instance, when a looming black circle was placed directly over their heads, they attempted to escape. They didn't react to all visual images, but still, the fact that they clearly demonstrated that their vision had been at least partially restored was unprecedented.

Even more surprising — because we had been taught it was impossible — was that the axons from the resurrected retinal ganglion cell grew all the way from the back of the eye to a place near the center of the brain called the optic chiasm.

"The amazing thing is that the regenerated axons were able to find their way home along these long, tortuous routes back to their targets in the brain . . . that's incredible," Huberman says.

NEURO GYM: HOW TO PARTICIPATE IN A CLINICAL TRIAL

The stem cell study I described at the beginning of this chapter could not have happened without people with glioblastoma who volunteered to participate. Several flew in from out of state in search of new medicine after their cancers returned. Clinical trials are the only way, in fact, that progress can be made in the treatment of diseases. Yet many people who could benefit don't, for all kinds of crazy reasons.

The craziest? Not wanting to offend your primary care doctor,

or the specialist he or she referred you to, by connecting with doctors running a clinical trial at a nearby university medical center. My
opinion: I am here to serve the patient as best as I can, and that means
sometimes referring them to a different specialist or academic center.
No doctor should ever take offense that you are going elsewhere for
treatment.

Another reason people don't seek out a clinical trial is because
they think they have to be extremely sick, looking for a last-ditch
treatment, in order to enroll. That's a myth. Even if you have only just
been diagnosed with a neurological disorder and have yet to receive
any treatment whatsoever, I encourage you to look into participating
in a clinical trial and at a minimum learning what is out there.

What's in it for you? First, you will receive state-of-the-art treatment from specialists. Second, even if the stem cells, drugs, or new
surgical approaches do not cure you, you get to help in the search
for treatments that do work. And third, the treatment will be entirely
free. Most important, the patient is always in the driver's seat. You
can always opt out anytime and go back to the treatments that exist.
Or choose no treatment at all.

So how do you begin? If you live in a rural area, your best bet is
probably to simply call or email the nearest university-affiliated hospitals and ask if they are running any clinical trials for your disorder.
Even if they're not, please consider heading over to get checked out
by a doctor there. They may offer guidance that can be implemented
closer to home by your local doctor.

If you have many university hospitals within driving distance
of your home or if you're willing to go anywhere, the best place to
start will be the websites of the National Institutes of Health or
other major medical organizations. Of course, you can always try
googling "clinical trial" and the name of the disorder you're battling. But here are some websites and phone numbers to get you
started:

NATIONAL CANCER INSTITUTE: https://www.cancer.gov/types/brain/clinical-trials. Telephone 1-800-4-CANCER.

NATIONAL INSTITUTE OF AGING studies of Alzheimer's and related disorders: https://www.nia.nih.gov/alzheimers/clinical-trials. Telephone 1-800-438-4380.

OTHER CLINICAL TRIALS BEING CONDUCTED BY NIH: https://clinicalcenter.nih.gov/participate1.html. Telephone 1-800-411-1222.

14

THE YOUNGER BRAIN

W hen I showed up to begin my undergraduate studies at the
University of California at Berkeley in August 1990, I was just
seventeen years old and ready to be dazzled by the intellectual firepower. Instead I found myself sitting in drab auditoriums with up to
800 other students listening to professors whose presentations were
videotaped so they could be shown again the next day to another 800
students, their words transcribed so you could read them without
even attending class. I grew so bored and disenchanted, I spent my
first two years mostly hanging with friends, drifting around Berkeley,
Oakland, and San Francisco. All the while, scoring Bs and Cs by cram-
ming the night before exams.

Rather than continue going through the motions, I decided in my
junior year to jump ship and drop out of college. My real interest
was to study human nature and that wasn't a course being offered.
So I rented a studio for $300 a month, took a job working as a secu-
rity guard at my old UC Berkeley cafeteria, and went looking for . . .
something.

Students who had sat next to me in class and who now saw me

wearing a blue uniform looked at me sideways. But I didn't care. Back then, my feeling was that what other people thought of me was none of my business. The job required only five hours a day, half at lunch and half at dinner. The rest of the time was mine.

Telegraph Avenue from Berkeley to Oakland became my classroom, where I studied the music shops, the restaurants, the galleries — and, best of all, the people. Hippies, bikers, blue-collar workers, tourists, professors, businesswomen, immigrants, children, and parents all went under my microscope. Still I wanted something more; I just didn't know what.

I heard that San Francisco General Hospital accepted volunteers. And there I had my first experience in a hospital: wiping blood off gurneys. This volunteer program was no candy striper gig. Here, volunteers were occasionally allowed to jump in along with technicians to do chest compressions if needed. That kind of access and exposure piqued my interest. It was my way of keeping an eye on something beyond living in the moment and life beyond working as a security guard.

At the time, medicine wasn't high on my list of career possibilities. The only doctors I had ever known were the general doctors my parents took me to who worked in office-like settings or clinics. Definitely important but nothing riveting.

I did respect that doctors helped people, and helping people was very much a priority for me. But it would need to be something exciting. Becoming a detective, a firefighter, or maybe a paramedic seemed interesting. But sitting around in an office writing prescriptions was not going to cut it for me.

Volunteering at SF General transformed my perception of medicine. I witnessed people at their rawest: crying, bleeding, choking, crashing, as well as enduring and surviving.

A year into my, um, sabbatical, the gurneys were all clean and lined up one night, so I had some spare time. I stood in the corner of the trauma bay as a patient was being "coded" in the center of the room by the nurses. A trauma surgery resident wandered over and stood next to me. She looked detached from the frenetic activity unfolding on the "crashing" patient, which left me a window to ask a question.

"Why don't they just shock the heart with those paddles?"

Leaning on the wall with her hands behind her and one leg folded to the side, as if she were just lazily waiting for a train, she said with a strange mixture of boredom and disdain: "You mean like on TV? That paddle-on-the-chest nonsense is only to shock them back into rhythm. This guy's heart is pumping but not strong enough. So if the drugs they are squeezing into him don't stop his blood pressure from bottoming out, I'm gonna slice between his ribs, stick my hand in his chest and squeeze the heart myself."

Whoa, I thought, that's medicine, too? Something called to me at that moment: a first glimpse of a profession I could devote my life to.

The next day, my job at the cafeteria felt as boring as the classes I had left behind. I went out and bought one of the many books I had faked my way through in high school. A few months later, as I was standing in my usual spot near the cafeteria exit reading *The Red Badge of Courage,* a female student walked by. We had never spoken but had noticed each other before.

"Oh, he reads," she said, her first words to me uttered with a twist of bemused sarcasm in her voice and the hint of a smile on her lips.

That would be the same woman to whom this book is dedicated: my wife.

Wandering the streets of Berkeley and working as a security guard was no longer enough. I had seen how intense medicine could be at SF General, and now here was a woman who inspired me and believed in me.

I tell this story of my days as a rebel without a clue because so many parents think their job is done once their kids head off to college. We all tend to focus on infancy and early childhood and the milestones that pediatricians keep track of. But the thing is, young adults *look* fully grown, and their intellectual powers — their capacity for making rapid calculations and forming long-term memories — will never get better.

But most young people reach their college years, like me, with plenty of growing still to do. Their decision-making and judgment lag behind. And that's not just because of a lack of experience. As neuroscientists have discovered, the brain — and in particular the prefrontal cortex — does not fully mature until our late twenties. Mine sure took a while.

NEURO GEEK: WHEN IS THE BRAIN FULLY GROWN?

Neuronal maturity doesn't happen at the same pace across the human brain. Some regions are completed much earlier than others, and the frontal lobes are the last to receive the finishing touches. The prefrontal cortex of the brain, the most complicated region, which provides us with cognition and judgment, demands the most sculpting. The final step in making a neuron work optimally is to have the long axonal cables wrapped in a fatty insulation from the surrounding glia, a process called myelination. Only then is the brain fully grown. Past studies based on examining brains after death were unclear, but the general understanding was this process occurred in late teens or early twenties. But with the widespread use of MRIs, a recent study reported remarkable findings. The researchers looked at myelination in 1,500 brains and showed that it occurred steadily into the twenties. And for some outliers even into their early thirties! This perhaps explains why some people seem to take longer than others to "find themselves" in a prolonged adolescence.

Now let's take a look at the epic story of a young brain's development, beginning at the beginning.

THE GREAT MIGRATION

How do you build a human brain? Given about forty weeks from the time of conception, nature decided eons ago to have a central spot for cranking out the billions of necessary neurons and then sending each of them on their way to migrate outward toward their designated permanent location. This radial migration is an astonishing feat of biological engineering.

Starting in the third week of gestation, a strip of stem cells, called the neural plate, forms along the back of the embryo. The neural plate folds outward to become the neural groove. Then the groove folds back together, like two ends of an omelet, to become the neural tube. The tube becomes the basis of the brain's fluid-filled ventricles, and from here, a stem cell factory on its banks begins kicking out baby brain cells at a rate of 250,000 *per minute*.

There are well over a dozen different types of these nascent neurons, some suited to vision, others to hearing or movement, others for interpreting signals from the body, and so on. Like herds in the savannah, the different types cling to each other and literally crawl together to their destined home, moving at about 60 MPH (that's *microns* per hour, or about one-five-hundredth of an inch). Most follow a path set by their predecessors, with each successive wave shoving and squeezing farther to build the brain from the inside out.

Neither the destinations nor the timing is haphazard; each neuron must arrive at an exact location at a predetermined time. Like FedEx packages, each neuron ends up delivered to its proper continent, country, county, city, street, and house number. Some, as it were, end up in Beijing; some go to London, some to Cape Town. And along the way to their eventual home, they send out thousands of synapses and dendrites to their neighbors. By birth, there are literally tens of trillions of synaptic connections between the neurons in an infant's brain.

But this great migration of neurons that builds the brain from the inside out is only the beginning of the story.

NEURO BUSTED: THE EVIL OF PHONES AND VIDEO GAMES

Greetings from the year 1954, when the book *Seduction of the Innocent* by psychiatrist Fredric Wetham became a bestseller. What was the great evil ruining children's minds back then? None other than the comic book. Batman and Robin were secretly encouraging homosexuality! Professor Marston's Wonder Woman was turning girls into lesbians! Monsters were filling the kids' heads with, um, monsters. It may seem like a joke now, but it was an absolute panic back then. Congress held hearings on this terrible vise, grilling William Grimes, publisher of *Tales from the Crypt* and *MAD*.

By the 1960s and '70s, it was television that was rotting the young mind. Apparently the *Brady Bunch* and *Love Boat* were corroding kids' brain cells.

Then came the first home video games in the early 1980s. Again, Congress held hearings. Parents' groups rose up to demand warning labels on excessively violent games.

And now we have the latest, most spectacular threat of all: the cell phone — vile robber of children's attention, purveyor of porn, instrument of peer pressure, destroyer of self-esteem, and ultimate time-waster.

Noticing a pattern here?

Of course cell phone use can and does get out of hand. Studies have linked excess phone use to tantrums in toddlers, to depression in teens, and to social isolation in young adults.

But, seriously, do you need a study to know that letting your three-year-old spend hours on a digital device is ridiculous? Do you need an "expert" to say that it's okay to tell your teen to put their phone down?

All I'm saying is that, in moderation, a cell phone is just another tool. With our kids we block porn. Other than that, we try to suggest a healthier digital diet. It's not just how much time they are on the phone, but what they are doing on it. No phones at dinner is a clear choice, and at meals we all put our phones down in a separate room, as studies show that having phones even nearby and face down on silent mode is still distracting. The title of one recent paper by researchers at the University of Texas and UC San Diego said it all: "The Mere Presence of One's Own Smartphone Reduces Available Cognitive Capacity." Their study of nearly 800 smartphone users found that those who were told to place their phones in another room performed better on a test of attention than did those allowed to keep the phones by their side. When I spoke with my teenage son about the study mentioned above, he wasn't surprised by the findings. He said, "Pops, its FOMO" (fear of missing out). Even when not directly holding a phone or gazing at the screen, there is a fear of missing out, so the siren call from the phone still tugs at one's attention.

But let's face it: cell phones are ubiquitous these days. My approach has been to set a few windows during the day when the phones are left behind in a different room. And as long as our kids are well otherwise, we try not to worry.

THE GREAT CULLING

The epic that is brain development involves not just growth, but death. During gestation, the fetal brain grows about twice as many neurons as it will eventually need. This surplus of juvenile neurons is culled and winnowed to what the maturing brain needs; the misfits that aren't picked to stay on the squad, so to speak, simply wither and die.

But do not mourn them. This dual process of creation and destruction is absolutely necessary for learning to take place. Synaptic pruning is the scientific term for how nature works like a master gardener, snipping stems here and there to permit other stems to grow and strengthen. You've heard the old line "Use it or lose it"? When it comes to synapses, that's how their fate is determined. Those assigned to discern visual input, store a memory, learn a skill, or control breathing get stronger; those without a job are left to wither.

Those neurons that do survive grow larger over the course of childhood, resulting in the brain's four-fold increase in overall size from birth until adulthood. The dendrites that stick around, meanwhile, slowly gain a fatty sheath covering them to improve their ability to conduct electrochemical signals.

Control of the pruning process is incredibly precise, and when it goes awry, neurodevelopment suffers. Children with autism, for instance, appear to have too many synaptic connections in certain parts of their brain early in life, and the usual culling that occurs in late adolescence is far less intense than that in children without the disorder. Perhaps, researchers believe, kick-starting the pruning process with a drug might help treat autism in the future.

Another serious brain disorder appears to be caused by the opposite problem: excessive pruning. People who carry genes that code for extreme culling of synapses are at increased risk for developing schizophrenia, marked by hallucinations and delusions.

But it's not all genetic. Nature *and* nurture both play an essential role. And that is where parenting comes in.

NEURO GYM: THE CARE AND FEEDING
OF A YOUNG BRAIN

Nobody should mistake me for a child-rearing expert. But I do know the neuroscience of what makes a developing brain healthy or diseased. And my wife and I have raised — are still raising — three sons, who are ages seventeen, fourteen, and thirteen.

As a scientist, surgeon, and dad, here are seven priorities if you're tending to a growing brain.

1. SAFETY FIRST. Working at Children's Hospital in San Diego, I saw the way children are most likely to die: through accidents. They drown in a pool, get hit by a car, burn in a fire, or fall from a window. All of these are ghastly — and preventable.

That's why my wife and I never allowed our kids to go into the street on our block when they were young, not even to visit a friend who lived on the other side. I don't care how many times you tell a seven-year-old to look both ways, they always forget. I built the fear of God into them about getting hit by an SUV.

Likewise, I had safety locks placed on all our windows above the ground floor so that they couldn't be opened beyond several inches. And as common as pools are in our part of southern California, we refused to have one in our backyard until all the boys were at least ten years old and could swim strongly.

2. ADVENTURE. The same parents on our block who thought we were overprotective for not letting our kids cross the street thought we were crazy for letting them play in the dry ravine behind our home. We had a five-foot-tall fence to keep out the coyotes, but on the other side were all sorts of critters and even snakes and the kind of weird, wild stuff that kids love. We allowed them to explore. They would throw rocks, play war games, look for

lizards, and otherwise screw around. I did keep a distant eye on them from the patio.

One day, when one of our sons was five, I watched as he jumped from the top of the fence to the ground below and landed badly. He started crying; I ran over, examined his leg, and could tell that he had a fracture or break. We drove him to Children's Hospital, and an X-ray showed that it was indeed a break. He had a little cast put on. Not a big deal.

Another time, a different son was riding his bike, trying to make it out underneath a closing garage door and missed the timing. So he got himself a nice scar on his forehead that is still there.

Even the designers of playgrounds have begun to wonder if maximizing their safety carries a cost. In response, so-called "adventure" playgrounds are sprouting up in Europe and the United States, where the usual swings and jungle gyms on thick padded surfaces are replaced with old tires, loose wooden beams, traffic cones, and even hammers and saws—all of it lying in the dirt. The Adventure Playground at the Berkeley Marina in California was one of the first to open in the United Sates, back in 1979. A newer one, called "The Yard," has drawn rave reviews since opening in 2016 on New York City's Governor's Island.

I'm all about safety, but kids need some risk in their lives to stimulate their minds, hone their individuality, and encourage their creativity. The era of helicopter parenting has gone too far. I let mine take a few bumps and knocks, free-range, and learn a little bit about risk from a young age.

3. TRANQUILITY. Schizophrenia, one of the most severe and often disabling mental disorders known, has a definite genetic component. But we also know that upbringing plays a role, in part because of a simple fact: it has become more common as people have moved into cities. The odds of developing the disorder, in fact, are

about twice as high in young people born and raised in cities compared to those in the country. Exactly what it is about urban life that increases the risk remains unproved, but the increased stress of urban life is hypothesized to be one of the reasons.

Similarly, children who have experienced severe trauma, whether emotional or physical, have three times the risk of developing schizophrenia as other children do.

What is the opposite of trauma and stress? It's peace and quiet, of course. That's why maintaining a calm home environment, no matter your circumstances, is important to your child's well-being. Easier said than done, I know.

4. SLEEP. As noted in chapter 6, getting enough sleep is essential to a growing brain. Most of us do a good job of hustling the kids to bed through elementary school, when we're in charge of their routines, but it's equally important in high school, when they are more in control of bedtimes. A recent study found that adolescents who don't get enough sleep tend to weigh more and have higher blood pressure and cholesterol. And those who routinely sleep less than seven hours per night tend to get lower grades in school than their IQ would predict.

One area of controversy that has come up regarding infants is the risks of co-sleeping: the sharing of a bed with your baby. The American Academy of Pediatrics recommends that infants sleep in the same room as their parents, because studies show that doing so reduces the risk of sudden infant death syndrome (SIDS) by as much as 50 percent. But they also recommend against letting an infant under the age of one share your bed and urge parents to never fall asleep with an infant on a couch, armchair, or other soft surface. Co-sleeping, studies have found, increases the risk of SIDS.

But other studies suggest that the real risk occurs when the parents use drugs or alcohol. And many parents believe co-sleeping enhances bonding, facilitates breastfeeding, and permits parents to get more sleep.

So when my wife and I were starting out as surgeons, working more than 100 hours per week, we chose to have each of our boys sleep in our bed with us for their first few years of life. We did what felt right. We wanted that physical and emotional connection with and for our kids. Did we go against medical advice? Yes. Am I suggesting you do the same? No. But even the American Academy of Pediatrics urges physicians to have open and nonjudgmental conversations with parents. Properly informed, parents can decide what they think is best.

5. ENRICHMENT. You have probably heard of the children raised in Eastern European orphanages who didn't get enough care and affection: Their brains were permanently injured by the lack of stimulation. They even had fewer folds on the surface of their brains. The simple fact is that babies and children require stimulation to grow normally. The less contact they get with interesting people, places, and activities, the more impoverished their brains will be.

My kids grew up with their grandparents, uncles, and aunts around much of the time. We thought having the grandparents watch them while we worked was out of necessity because of our crazy schedules. But in retrospect it was Miracle-Gro for their brains. Ours was a very old-world approach, commune-style, with nonstop chattering in Spanish and English, lots of hugs and walks around the neighborhood.

When it was time for them to go to preschool, we went for those that were most attentive to them emotionally, not academically. Learning would come later, much later — at that point they just needed the warmth. We actually ended up placing them in a Hebrew preschool, despite the fact that neither my wife nor I was Jewish, simply because we saw the teachers always holding the little ones in their arms or laps whenever we visited.

Enrichment, however, is not just emotional or intellectual. As the boys have grown older, we have encouraged them to participate in a variety of sports and physical activities, while discouraging them from overspecializing. These days, too many teens are doing no exercise at all — or, on the other extreme, devoting themselves to a single sport as if they were professionals. Travel leagues have kids and their parents spending their weekends driving to and from games. If that's your thing and your kid is loving it, good for you. But if you're feeling pressured to get your kid into a travel league because "everybody's doing it," think twice. Relentless focus on a single sport can cut kids off from new experiences and from kids whose parents can't afford the time or money to participate.

Our sons are on the baseball teams at their school and sometimes participate in a fall league. But the rest of the year, we just play stickball in the backyard and throw the tennis ball around (including with our nondominant hand to improve coordination).

In fact, when summer comes around, we make a point of *not* enrolling our boys in any athletic leagues or sending them to camp.

"But Dr. Jandial," you might ask, "isn't summer camp an enriching experience where children meet other kids and engage in a variety of healthy outdoor activities?"

Absolutely. And if you want to send your kids to camp, go for it. But my feeling is that sometimes even boredom can be enriching for kids, at least when most of their year is so structured. I'm unconventional in this regard, but I consider summer to be an excellent time for kids to get bored enough that they have to invent their own pastimes.

All things in moderation, even enrichment. Sometimes kids have to just mess around. It's part of growing up.

6. NUTRITION. Contrary to many parents' beliefs, most children do not need to be given vitamin supplements. Although the $30 billion-a-year supplement industry would have you think otherwise, no good evidence exists to show that otherwise healthy kids *or* adults benefit from taking vitamin supplements. An important exception: pregnant women should take folate to prevent neural tube defects, as well as iron and calcium supplements. In addition, breastfed infants should be given a vitamin D supplement until they are weaned, and all infants should be given an iron supplement between the ages of four and six months. Other than that, according to a review of prior research by Harvard researchers published in 2018 in the *Journal of the American Medical Association*: "Healthy children consuming a well-balanced diet do not need multivitamin/multimineral supplements, and they should avoid those containing micronutrient doses that exceed the RDA [Recommended Dietary Allowance]. In recent years, omega-3 fatty acid supplementation has been viewed as a potential strategy for reducing the risk of autism spectrum disorder or attention-deficit hyperactivity disorder in children, but evidence from large randomized trials is lacking."

Another thing you might consider eliminating is soda — and not just the sweetened varieties. Diet soda was found in a study by researchers at UC-San Diego to shift the brain's taste preference toward sweets, increasing the risk of obesity.

You might be surprised to learn that fruit juice is almost as bad as soda. In fact, a 12-ounce glass of orange juice contains nearly as much sugar as a can of Coke. As three leading pediatricians recently wrote in the *New York Times*: "More than half of preschool-age children (ages 2 to 5) drink juice regularly, a proportion that, unlike for sodas, has not budged in recent decades. These children consume on average 10 ounces per day, more than twice the amount recommended by the American Academy of Pediatrics. . . . In the past decade or so, we have succeeded in recognizing the harms of sugary beverages like soda. We can't keep pretending that juice is different."

7. PERSISTENCE. I began this chapter with the story of my unconventional journey through college. Like all other young adults, my brain had not yet finished maturing. The emotions and desires of the limbic system, deep in the brain's depths, had yet to be tamed by the patience, planning, and control of the frontal lobes.

That's why, as a parent, I am determined to remain as close as possible to my three sons as they enter their teen years and beyond — and I encourage you to do the same. Sure, they need independence, but modern neuroscience shows why they still need close attention past high school: the brain's finishing touches happen later than we previously thought. Hanging out with them regularly is one of the best ways to stay close.

My routine is to tell our boys that they can do what they want during the day and later in the evening, but that I need them for an hour around dinnertime. We might go for a swim, do some pull-ups, work out with free weights, or play video games together. We grill some salmon and boil some broccoli. And I try every night to have us all sit down and read something then talk about it. I might ask them to read a news article, or we'll read aloud a story from one of the many magazines I keep around the house, like *Wired* or *Rolling Stone*. My philosophy is that they get fifteen hours to do what they want the rest of the day; they can give me sixty minutes.

Even when they leave for college, keep in mind that only half of all students who enter college will graduate. So keep on texting and calling — and visiting them in person. Their developing brains still need you. And if you are so unlucky as to have a rebellious kid like me who marches to the beat of his own drum and insists on doing it his way, just remember: you might have a budding brain surgeon on your hands!

15

THE OLDER BRAIN

D r. Bernstein, did you fall down a few weeks ago?"

"Please don't call me doctor. I'm just William; I'm not a physician today. What did the CT scan show?"

He was ninety-one, an age that fewer than one in six men ever reach. Even then, one-third of the men who *do* reach their nineties already have dementia. But William was one of those rare old-timers: not merely conversant and functional, but witty, curious, sharp, and totally up to date on medical advancements, even though he had retired a decade earlier. He was so sharp that I skipped the usual mini mental exam we do as part of our history and physical. Bald, trim, and with a strong white goatee, he also kept up his appearance aside from a thicket of hair in his ears.

"Well, sir," I told him, "the CT shows a subdural hematoma. It's substantial and threatening."

"There you go again with the excessively deferential stuff. Don't call me sir. I need you to be my surgeon and not treat me any differently."

"William, you're twice my age, so I was just being respectful."

"Actually, I'm closer to three times your age."

According to his chart, William had come into the hospital around 8 a.m. that morning reporting difficulty moving his left arm. The CT scan showed a massive bleed in the right hemisphere in the space between his brain and the dura, the brain's sheath. That's what is meant by a *hematoma* (bleed) that is *subdural*.

"How new?" he asked.

"Not very," I said.

On the scan, the blood showed up as a dark silhouette, which meant it was old blood, approximately a few weeks. Fresh blood shows up as bright white on a CT scan.

"How big?" he asked.

"Sizable," I said. "About two inches thick and seven inches wide, covering your entire right hemisphere."

"Damn aspirin," he said. "Probably caused the bleed."

It didn't cause the bleed, but it interfered with the body's own efforts to close off the torn veins. By thinning the blood, a baby aspirin per day significantly lowers the risk of heart attacks and strokes due to blood clots. But that same clot-busting ability also increases the risk of bleeding in the stomach and in the brain. Overall, far more lives are saved by taking the baby aspirin, because the risk from a heart attack or stroke due to a clot is so much higher than the risk of bleeding. But that wasn't much solace to William right now.

"We'll give you vitamin K to reverse the blood thinning effect of the aspirin. And I'll have you stay overnight. In the morning we'll get you a second scan and see how you're doing."

The next morning I walked into his room, and he said, "It's growing."

"Let's see what today's brain scan shows," I replied.

"I know what it shows," he said. "My left arm is worse."

He was right. The scan showed that the clot had grown.

"No offense," I said, "but you have significant brain volume loss and that has probably saved your life so far."

"Some offense taken," he said with a smile.

He knew I wasn't joking. On average, a person's brain volume shrinks by about 5 percent per decade after the age of forty, even though the size of the dura surrounding it (and, of course, the skull) remain the same. At age ninety-one, William's brain was two inches below the dura instead of flush up against it, as in a young adult. That two inches, however, had given the blood room to collect before it began exerting harmful pressure on the brain — until yesterday morning, when he began to have difficulty moving his left arm.

"So when are you scheduling surgery?" he asked.

"You're ninety-one. I'm not thinking surgery is the way to go. We have medications to lower your blood pressure and —"

"Bullshit," he said. "We're not talking about a rickety knee. Even the weak arm is no big deal. I can get by with my right hand. I'm worried about my mind. Give that hematoma an arm and next it'll take my memory. So, tell me about my surgical options, Dr. Jandial."

The next morning, under the surgical drapes, his flesh belied the brilliant mind underneath. The skin on the scalp was paper thin, like eyelid skin, and I sliced more softly. The subdermal layer of fat that gives our scalps that fleshy feel was no longer neon yellow but now looked sun-faded. The skull no longer was bright ivory but the beige hue of an old manila envelope. Time had ossified the natural soft bone on the inside of the skull, so I had to drill harder and longer. The dura had also become calcified and adherent to the inside of the skull; it was no longer the lithe, slippery sheath it once was.

With two holes drilled into his skull, the dura was the last layer to open. I took a long, thin scalpel with an X-Acto blade–like tip in my right hand and touched the back end of the scalpel with an electrical cautery instrument called a Bovie. The current ran through the stem of my scalpel and tip and cut and singed the dura in one move. My latex glove kept me safe from the current, and the blood clot kept his brain safe from the current. Next came the satisfying part: old blood, which looks like motor oil, sprayed into the air over my right forearm. His brain had indeed been under a lot of pressure.

With much of the gunk gone, I used my surgical loupes to look through the opening in the dura. The gap between brain and dura, I saw, was not empty: Veins stretched from the brain to the skull's underside, like the strings of a marionette. And one of them had a drop of fresh blood oozing from a tear. I singed it shut.

The next step was to clear out all the remaining old blood. To do it, I used a highly specialized piece of advanced technology: a turkey baster. I filled it with sterile water, squirted it in one of the holes in the skull, and the dark motor oil–like fluid squirted out the other hole. Then I moved to the opposite hole and repeated. Back and forth until the fluid that was being lavaged out went from dark brown, to red, to pink, to clear. I had washed off the surface of his brain.

As the light shone from my headlight into one of the holes I had made, I peered in and could see the surface of the collapsed brain, like the curved surface of the earth rotating away from me. A window to its hemispheric curvature over multiple lobes. The ridges were not opalescent, as in youth. Time had built up the residue of aging, called gliosis, and the color was now yellowy tan with dark spots from small bleeds. The diaphanous arachnoid covering of the brain had become dusky like a cataract. The surface of his brain had the appearance of a deserted planet.

But appearances can be deceiving. Later that afternoon, after William had awakened, I stopped by to see him in the neuro ICU.

"How'd it go?" he asked.

"As it should," I said.

"I know," he said, and lifted his left arm to show me that he had recovered movement there.

Although his brain had lost its youthful sheen, his mind remained razor sharp, teaching us that brain atrophy does not equal mind atrophy.

NEURO GEEK: SPIKE THE NOURISHING LIQUOR IN YOUR BRAIN

Blood isn't the only liquid circulating inside you. Our brains are bathed in cerebrospinal fluid (CSF), called the "nourishing liquor" because it's rife with growth factors that keep your neurons fertilized. In fact, neurons grown in a petri dish in the laboratory shrivel up and die if the aqueous environment isn't replete with growth factors. With normal aging, these growth factors (called neurotrophins, e.g., BDNF, NGF, CTNF, GDNF) are less abundant in the brain milieu, and, as a result, neurons can become sluggish and even undergo cell death. However, there is a way to stave off this loss of nutrients in your CSF and even replenish them to youthful levels: exercise. Specifically, an exercise regimen with *both* aerobic exercise and resistance training has been shown to be the best way to keep your brain's nourishing liquor at full strength.

TYPICAL BRAIN AGING

Brain shrinkage is just one of many processes that occur in the older brain, some of which, as we shall see, actually bring remarkable benefits. The most obvious change is with certain kinds of memory. There are actually four kinds of memory:

SEMANTIC MEMORY is general knowledge about the world, everything from who Isaac Newton was, to what a bagel tastes like, and where your office is — all the kinds of basic facts and meanings that computers and robots don't understand. Thankfully, in normal brain aging, this huge knowledge base not only remains generally stable in an older adult but can continue to grow as a person learns more.

PROCEDURAL MEMORY is knowledge of *how* to do things. How do you get dressed in the morning? How do you ride a bike? Once learned, procedural memories tend to remain solid as a rock. Performance does tend to fall off with age due to the natural slowing of reflexes, although as of this writing, seventy-six-year-old Morgan Shepherd is still competing as a NASCAR driver.

EPISODIC MEMORY is your recollection of events. Where did you go to kindergarten? When did you first meet your spouse? What did you eat for breakfast yesterday? And where did you leave the keys? This is the one that tends to naturally weaken with age. In fact, episodic memory peaks in your midtwenties and then slowly declines throughout life. That's why you still remember the words to songs from when you were a teenager but barely remember the plot of a movie you saw last year. Fortunately, our phones can help catalog and retrieve much of this type of memory.

WORKING MEMORY is the brain's workspace, where you can hold and manipulate a handful of facts and figures in your head. The reason multiplying 36 by 42 in your head is so difficult is that your working memory has limits — and the older you get, the tighter your limits. That's why mathematicians, musicians, and physicists tend to do their most important work when they're young . . . and why once-simple tasks get progressively more difficult with age. This the type of memory, which facilitates multitasking and juggling life's responsibilities, is the type of memory most healthy people want to optimize. Working memory plus creativity is the key to productivity. Brain-training companies and programs are paying a lot of attention to keeping working memory at peak levels.

What are the physical changes underlying these changes in memory and other brain functions? There are at least four of them:

1. **LOSS OF SENSORY FUNCTIONS**, in particular due to difficulties
    seeing and hearing. Studies have shown that age-related hearing
    loss is directly associated with cognitive decline, in part because
    the area of the brain that is supposed to be dedicated to higher-
    level cognition is instead forced to struggle to interpret dimin-
    ished sounds.
2. **HAROLD, OR HEMISPHERIC ASYMMETRY REDUCTION** in older
    adults. Rather than showing the normal variation in electrical
    activity between the left and right prefrontal cortex as seen in
    younger people, the two sides of the brain tend to show increas-
    ingly similar activation when faced with tasks involving memory
    or visual perception. It seems to be a case of "all hands on deck,"
    because tasks that previously could be handled easily by one side
    of the brain now need help from the other.
3. **REDUCTION IN NEUROTRANSMITTERS.** Beginning in early adult-
    hood, dopamine levels, which affect both physical movement
    and reward-motivated behavior (in addition to a bunch of other
    things), decline by about 10 percent every decade. The result for
    some people who had naturally low levels of dopamine to begin
    with is that they develop Parkinson's disease, in which the mus-
    cles become increasingly stiff, slow, and shaky. For others, it's the
    fever of youthful ambition that slowly fades. Levels of serotonin
    and other neurotransmitters likewise drop with aging.
4. **HORMONE LEVELS** also drop with age in both men and women.
    Early in life, growth hormone, estrogen, and male steroids all
    play essential roles in brain structure and function, and their
    gradual decrease over the lifespan may be associated with
    changes.

But not all the brain changes in older people are bad. Research has
found that older adults have a greater sense of well-being and greater
emotional stability than do teens or young adults. Dilip Jeste, MD,

director of the Sam and Rose Stein Institute for Research on Aging at UC San Diego, has written a series of papers in prominent psychiatric and neurologic journals exploring what he calls the neurobiology of wisdom. "As neuroscience advances," he says, "it's taken on things once dismissed as fuzzy, like consciousness, and made them into legitimate areas of research. Wisdom is the inevitable final subject, the one that scientists were too afraid to deal with until now." One of his most intriguing findings, he says, is that as people age, "Physical health goes down, cognitive abilities go down, but happiness increases with age, satisfaction with life increases with age. That's due in part to the growth of wisdom about what really matters."

NEURO BUSTED: SOCIAL MEDIA DOESN'T COUNT AS SOCIALIZING

An interesting study found that super-agers tend to have stronger social networks on average than people whose cognitive performance declines normally. Senior citizens are the fastest growing demographic on some social media sites, but whether social media and online interactions count as "socializing" from the point of brain aging is just beginning to be investigated. The early reports are promising. One recent study showed that older people who have come on board with social media and technology have fewer chronic illnesses and depressive symptoms. Another study from the *Proceedings from the Royal Society* even showed that the density of neurons in the temporal lobes, which serve language and memory, is greater in those with online social networks.

As with kids and younger adults, too much time spent on social media can be depressing or anxiety-producing. All things in moderation. But the greater danger, by far, is when an older person becomes isolated and disconnected. So by all means, older adults can

and should be encouraged to connect online. I'm confident studies
will show that social media serves old folks well.

HEART AND MIND

One of the key reasons that rates of dementia have fallen sharply since
the 1970s is the advent of improved treatments for heart ailments.
What's good for the heart is actually very good for the brain. The
steps you take to keep your heart arteries unclogged also keep brain
arteries open. Cholesterol-lowering drugs have dramatically reduced
coronary artery disease and are effective even in people who live sed-
entary lifestyles and eat foods that aren't "heart healthy." Statins, prescribed to lower cholesterol, have lately been shown to lower the risk
of Alzheimer's disease in most people. A giant study of 400,000 Medi-
care beneficiaries recently found that men who took statins had a 12
percent reduced risk of Alzheimer's on average, while women had a
15 percent reduced risk.

High blood pressure has long been known to increase the risk of
cognitive impairment and Alzheimer's disease. For every 10 mmHg
increase in systolic blood pressure, one large study has found, the
risk of poor brain health goes up by 9 percent. Because one in three
American adults has high blood pressure, it's important to get yours
checked and the sooner the better: high blood pressure in midlife is
strongly linked to the risk of developing Alzheimer's later in life.

Avoiding blood clots by taking a blood thinner can also benefit
the brain, but the drugs also carry some risks. The fact that William
took a baby aspirin per day, for instance, might have contributed to
the bleeding in his brain that required surgery to repair. Only people
diagnosed with heart disease are generally recommended to take a
baby aspirin. Even then, in people over seventy-five who take it after
a stroke or heart attack, new research suggests that their risk of developing a potentially fatal stomach bleed is higher than previously

thought. So even though aspirin is sold over the counter, keep in mind that it's serious medicine. See a doctor before taking it.

Finally, another medical condition known to seriously increase the risk of heart disease is now also known to wreak havoc with the aging brain. Diabetes — in particular the persistently high blood sugar levels of poorly treated diabetes — substantially raises the risk of dementia. And according to a recent study of nearly 13,000 older adults, the higher their blood sugar levels tended to peak, the greater their risk of developing dementia.

NEURO GYM: SECRETS OF HEALTHY BRAIN AGING

There is good news about the aging brain: more people than ever are living into their eighties and beyond in excellent cognitive health as the rate of Alzheimer's plummets. So, even though the total number of people with dementia is increasing because people are living longer, the risk of getting dementia is actually lower. Since the 1970s, in fact, the risk of dementia due to any cause has fallen by 20 percent *every decade,* proving that lifestyle plays a major role in how our brains age, and that dementia is not an immutable force.

How can you be a super-ager like William? There are three ways to increase your odds:

EDUCATION. In people with at least a high school education, the rate of dementia has fallen by nearly half since the 1970s. Numerous studies have shown that education plays an important role in reducing the risk of later developing dementia. People with college degrees and beyond, like William, do better on average than those with high school degrees, but even those who simply started college but didn't finish it tend to remain cognitively healthy longer, on average, than people who never entered. And high school graduates do better, on average, than dropouts.

The reason why education pays off is because of something called cognitive reserve: people with extra brain power (thanks in part to extra education) can afford to lose more before showing obvious signs of decline. That's why two people who have brains that look exactly alike — with the same amount of shrinkage — can nevertheless show dramatic differences in how long they remain cognitively healthy. Those who put their brains to better use can withstand greater loss of brain matter.

SOCIAL NETWORKS. Neuroscientist Emily Rogalski runs the Super Agers study at Northwestern University in Chicago, where she follows a group of two dozen people aged eighty and over who retain the cognitive faculties of people in their fifties. To select recruits, she reads them a list of fifteen random words and then asks them to remember them a half hour later. The average eighty-year-old remembers only five. The average fifty-year-old remembers nine. Her super-agers remembered at least nine, and some remembered all fifteen words!

The one factor that distinguished her super-agers from others is that they were more extroverted and had more social contacts than their peers. That shouldn't be surprising: multiple studies have shown that the greater your degree of social connections, whether with friends or family, the lower your risk of developing dementia. Compared with those who remain in touch with few friends or family, studies have found, those with many relationships have a risk of dementia that is between 25 percent and 50 percent lower.

That doesn't mean you have to turn into a social butterfly overnight at the age of seventy-five. It does mean, however, that getting out of the house and connecting with people has real physical benefits for your brain, which requires stimulation to maintain its health.

Then again, social isolation is not the same as loneliness. Some studies have found that it's loneliness — the *feeling* of being isolated — that puts people at higher risk of cognitive loss. For those who feel perfectly happy with a book and a cup of tea, more power to them. But if you ache for more connections, reach out.

However, President Ronald Reagan had far more social interactions than most of us could ever dream of, yet he was diagnosed with Alzheimer's disease at the age of eighty-three. So we are talking about averages here. Don't let this talk of averages scare you. Loners are not destined to develop dementia, just as social butterflies aren't immune from its depredations.

PHYSICAL ACTIVITY. Physical activity turns out to be one of the absolute best ways to maintain and even improve cognitive health.

Countless studies have shown that exercise programs directly contribute to brain function. Surprisingly, though, cardiovascular exercise has demonstrated the least benefit in older adults. By contrast, Teresa Liu-Ambrose, PhD, of the University of British Columbia, has found that resistance training with weights improves cognition. Even more surprisingly, at least to me, is that similar benefits on cognition were seen in two clinical trials — one in the *Journal of Alzheimer's Diseases,* another in the *American Journal of Geriatric Psychiatry* — among older adults randomized to participate in the traditional Chinese practice of tai chi.

EPILOGUE

I have tried to share the thrill and joy of my life's work in the brain and the pride I feel in the progress we have made in not only treating its disorders but enhancing its function and marveling at its complexity.

Our understanding of the brain's structure has progressed exponentially in the past few decades, giving us new insights into how memories are encoded and language embodied; how creativity arises and can be boosted; and how essential sleep and deep breathing are to health. We've learned the truth about concussions, "smart" drugs, and so-called brain foods. We have discovered undeniable evidence of brain plasticity and the brain's ability to heal and reorganize itself in the face of injuries and trauma. Sci-fi notions of implants and brain-machine interfaces have become reality. Stem cells and the body's own immune system are being harnessed as powerful new weapons against disease. Dementia rates are dropping while the number of super-agers reaching their nineties in good cognitive health is rising. And the discovery that children's brains continue maturing until their thirties has brought relief and understanding to countless parents who are worried that their kids are still, like I was, struggling in early adulthood.

Yes, there is much cause for celebration in neuroscience. But being a brain surgeon who operates routinely on tumors also involves no small share of disappointment and tragedy. My patients remind me that there is so much more we have yet to learn. And this gives me hope that within my lifetime and yours, through continued exploration into the inner workings of our brains, breakthroughs will come.

For now, I hope I have debunked some of the pop neuroscience myths with the Neuro Busted sections of this book, that you have learned some of the colorful aspects of neuroscience in the Neuro Geek sections, and that you will take advantage of the practical tips I have presented in the Neuro Gym sections.

I hope I have conveyed to you the single most important discovery that neuroscientists have made since I became a brain surgeon: your brain's health is within your own control. Your willingness to engage in lifelong learning, social engagement, and a childlike openness to new experiences will define your brain's fate.

In neurobiology, fitness is a major concept. Under stress, some cells will outperform their neighbors — a molecular survival of the fittest. Similarly, neurofitness borrows from that concept and extends it to each unique brain navigating an evolving world with challenging experiences and environments. Throughout the last century and even now, the emphasis has been mostly on heart and physical fitness. My hope is that this book has shown you that today, tomorrow, and from here on, the focus should be on the fitness that matters the most — NEUROFITNESS!

ACKNOWLEDGMENTS

D eb Brody from HMH gave me the chance to publish this book. Then she allowed me the room to wander creatively and share some of my patients' stories and a few of my own. For this, I am grateful. Her team, including Olivia Bartz, Allison Chi, and Rebecca Springer, has been gracious with their time and energy. It is recognized and appreciated.

Dan Hurley is a brilliant collaborator who helped distill and deliver the best of what I had to share. Annie Hurley is a talented artist and made the book better with her art. Along the way, I had invaluable conversations, often while driving the freeways of Los Angeles, with my dear friend Sam Hughes. It doesn't hurt that he is a neurosurgeon who also has a PhD in classics.

Ryan McNeily and Amanda Kogan from WME television connected me with their literary colleague, Mel Berger. This book and the fulfillment of my dream of becoming an author would not have happened if Mel hadn't let me into his ecosystem. Then he went further and mentored me on how to understand and approach the literary landscape, gently trimming some of my wayward sails and letting others unexpectedly loose. For this, I am grateful and indebted.

As always, I must thank my patients for allowing me to learn from the depth and complexity of their journeys.

NOTES

page

PROLOGUE

4 *without demonstrating any attributable mental dysfunction:* JD Handley, DM Williams, JW Stephens, et al., "Changes in Cognitive Function Following Bariatric Surgery: A Systematic Review," *Obesity Surgery* 26, no. 10 (2016): 2530–6.

mind-calming effects of meditative breathing: GN Levine, RA Lange, CN Bairey-Merz, et al., "Meditation and Cardiovascular Risk Reduction: A Scientific Statement from the American Heart Association," *Journal of the American Heart Association* 6, no. 10 (2017): doi: 10.1161/JAHA.117.002218

1. AN ANATOMY LESSON LIKE NO OTHER

23 *Four slices of Einstein's brain:* MC Diamond, AB Scheibel, GM Murphy Jr, T Harvey, "On the Brain of a Scientist: Albert Einstein," *Experimental Neurology* 88, no. 1 (1985): 198–204.

2. BEYOND MEMORY AND IQ

26 *Flynn made a curious discovery:* JR Flynn, "The Mean IQ of Americans: Massive Gains 1932 to 1978," *Psychological Bulletin* 95, no. 1 (1984): 29–51.

27 *Flynn has written:* JR Flynn, "Are We Really Getting Smarter?"
    Wall Street Journal, September 21, 2012. https://www.wsj.com/
    articles/SB10000872396390444032404578006612858486012

28 *paper published in 1950 by psychologist Karl Lashley:* K Lash-
    ley, "In Search of the Engram," *Society of Experimental Biology,*
    Symposium 4 (1950): 454–82.

    *Thompson found that if he removed just a few hundred neu-
    rons:* GA Clark, DA McCormick, DG Lavond, RF Thompson, "Ef-
    fects of Lesions of Cerebellar Nuclei on Conditioned Behavioral
    and Hippocampal Neuronal Responses," *Brain Research* 291, no. 1
    (1984): 125–36.

    a picture of Jennifer Aniston: R Quian Quiroga, L Reddy,
    G Kreiman, et al., "Invariant Visual Representation by Single
    Neurons in the Human Brain," *Nature* 435 (2005): 1102–7.

29 *area-restricted search:* TT Hills, R Dukas, "The Evolution of
    Cognitive Search," *Cognitive Search: Evolution, Algorithms and
    the Brain* (Cambridge, MA: The MIT Press, 2012): 13.

30 *a really cool study:* N Unsworth, GA Brewer, et al., "Work-
    ing Memory Capacity and Retrieval from Long-Term Memory:
    The Role of Controlled Search," *Memory and Cognition* 41, no. 2
    (2013): 242–54.

31 *an amazing series of experiments:* M Gagliano, M Renton,
    M Depczynski, et al., "Experience Teaches Plants to Learn Faster
    and Forget Slower in Environments Where It Matters," *Ecologia*
    175, no. 1 (2014): 63–72.

    an even more astonishing report: M Gagliano, VV Vyazovskiy,
    AA Borbely, et al., "Learning by Association in Plants," *Scientific
    Reports* 6 (2016): 38427. doi: 10.1038/srep38427

32 *was fined $2 million:* "Lumosity to Pay $2 Million to Settle FTC
    Deceptive Advertising Charges for Its 'Brain Training' Program."
    https://www.ftc.gov/news-events/press-releases/2016/01/
    lumosity-pay-2-million-settle-ftc-deceptive-advertising-charges

33 *risk of developing dementia nearly cut in half:* D Hurley, "Could Brain Training Prevent Dementia?" *The New Yorker*, July 24, 2016. https://www.newyorker.com/tech/annals-of-technology/could-brain-training-prevent-dementia

34 *brain game called Robot Factory:* AK Brem, JN Almquist, K Mansfield, et al., "Modulating Fluid Intelligence Performance Through Combined Cognitive Training and Brain Stimulation," *Neuropsychologia* 118, pt. A (2018): 107–14. doi: 10.1016/j.neuropsychologia.2018.04.008
    As science journalist Daniel Goleman showed: D Goleman, *Emotional Intelligence: Why It Can Matter More Than IQ* (New York: Bantam Books, 1995).

35 *Grit and determination:* A Duckworth, *Grit: The Power of Passion and Perseverance* (New York: Scribner, 2016).
    has published studies: KA Ericsson, WEG Chase, et al., "Acquisition of a Memory Skill," *Science* 208 (1980): 1181–82.

36 *In his book* Outliers: M Gladwell, *Outliers: The Story of Success* (Boston: Little, Brown and Company, 2008).

37 *self-testing enhances learning:* HL Roediger, JD Karpicke, "Test-Enhanced Learning: Taking Memory Tests Improves Long-Term Retention," *Psychological Science* 17, no. 3 (2006): 249–55.

3. THE SEAT OF LANGUAGE

41 *Broca conducted an autopsy:* PP Broca, "Loss of Speech, Chronic Softening and Partial Destruction of the Anterior Left Lobe of the Brain," *Bulletin de la Société Anthropologique*, 2 (1861): 235–38.

42 *an injury to Wernicke's area:* C Wernicke, "The Symptom-Complex of Aphasia," *Diseases of the Nervous System* (1908): 265–324.
    English-only speakers performed slower: B Zhou, A Krott, "Bilingualism Enhances Attentional Control in Non-Verbal Con-

flict Tasks — Evidence from Ex-Gaussian Analyses." *Bilingualism: Language and Cognition* 21, no. 1 (2018): 162–80.

43 *dual-language kids had gained a full year:* P Cornwell, "Study: Students in Dual-Language Programs Outperform Peers in Reading," *Seattle Times*, November 18, 2015. https://www.seattletimes .com/education-lab/in-portland-dual-language-students-out perform-peers-in-reading/

study published in 2007: E Bialystok, FL Craik, M Freedman, "Bilingualism As a Protection Against the Onset of Symptoms of Dementia," *Neuropsychologia* 45, no. 2 (January 28, 2007): 459–64.

As a recent review: D Perani, J Abutalebi, "Bilingualism, Dementia, Cognitive and Neural Reserve," *Current Opinion in Neurology* 28, no. 6 (December 2015): 618–25.

4. UNLEASH CREATIVITY

55 *researchers from the University of Utah:* JA Nielsen, G Brandon, A Zielinski, MA Ferguson, et al., "An Evaluation of the Left-Brain vs. Right-Brain Hypothesis with Resting State Functional Connectivity Magnetic Resonance Imaging," *PLOS ONE* 8, no. 8 (August 14, 2013). https://doi.org/10.1371/journal.pone.0071275

57 *an excellent documentary: I Remember Better When I Paint: Treating Alzheimer's Through the Creative Arts,* directed by E Ellena and B Huebner (2009).

59 *Salvador Dali wrote:* S Dali, *50 Secrets of Magic Craftsmanship* (New York: Dial Press, 1948).

60 *study by Eric Schumacher:* A Griffin, "People Who Daydream Are More Intelligent, and May Get Distracted Because They Have 'Too Much Brain Capacity,'" *The Independent,* October 25, 2017. https://www.independent.co.uk/news/science/day dream-intelligence-smart-study-lost-in-thought-meetings-mri-research-a8019391.html

61 *Play teaches them some lifelong skills:* Case Western Reserve University, "Psychologist Explores How Imaginary Play in Childhood Stirs Creativity," *The Daily,* November 13, 2013. https://thedaily.case.edu/case-western-reserve-university-psychologist-explores-imaginary-play-life-adult-creativity-in-new-book/

63 *a test of creativity:* RA Atchley, DL Strayer, P Atchley, "Creativity in the Wild: Improving Creative Reasoning Through Immersion in Natural Settings," *PLOS ONE* 7, no. 12 (2012). doi: 10.1371/journal.pone.0051474

outdoor play or sports: C O'Mara, "Kids Do Not Spend Nearly Enough Time Outside," *Washington Post,* May 29, 2018. https://www.washingtonpost.com/news/parenting/wp/2018/05/30/kids-dont-spend-nearly-enough-time-outside-heres-how-and-why-to-change-that/

64 *perhaps the weirdest human ever to win the Nobel Prize for chemistry:* P Carlson, "Nobel Chemist Kary Mullis, Making Waves As a Mind Surfer," *Washington Post,* November 3, 1998. https://www.washingtonpost.com/archive/lifestyle/1998/11/03/nobel-chemist-kary-mullis-making-waves-as-a-mind-surfer/

65 *Doing LSD was one of the two or three most important things:* W Isaacson, *Steve Jobs* (New York: Simon & Schuster, 2011).

5. SMART DRUGS, STUPID DRUGS

68 *A recent online survey:* B Maher, "Poll Results: Look Who's Doping," *Nature* 452 (2008): 674–75. https://www.nature.com/news/2008/080409/full/452674a.html

14 percent said they had used a stimulant: A Frood, "Use of 'Smart Drugs' on the Rise," *Scientific American,* July 5, 2018. https://www.scientificamerican.com/article/use-of-ldquo-smart-drugs-rdquo-on-the-rise/

69 *88,000 people die each year:* Centers for Disease Control and

Prevention: "Alcohol Fact Sheet." https://www.cdc.gov/alcohol/
fact-sheets/alcohol-use.htm

Alcohol appears to modestly reduce the risk: Harvard T.H. Chan
School of Public Health, "Alcohol: Balancing Risks and Benefits." https://www.hsph.harvard.edu/nutritionsource/healthy-
drinks/drinks-to-consume-in-moderation/alcohol-full-story
/#possible_health_benefits

a few drinks can actually improve people's ability: AF Jarosz,
GJ Colflesh, J Wiley, "Uncorking the Muse: Alcohol Intoxication
Facilitates Creative Problem Solving," *Consciousness and Cogni-
tion* 21, no. 1 (2012). doi: 10.1016/j.concog.2012.01.002

70 *an article published in the April 2015 issue of the* Atlantic: G Glaser, "The Irrationality of Alcoholics Anonymous," *Atlantic,* April
2015. https://www.theatlantic.com/magazine/archive/2015/04/
the-irrationality-of-alcoholics-anonymous/386255/

less than 10 percent of people who enter AA achieve sobriety:
L Dodes, *The Sober Truth: Debunking Bad Science Behind 12-Step
Programs and the Rehab Industry* (Boston: Beacon Press, 2014).

NPR's RadioLab *has aired a compelling segment: Radiolab,*
"The Fix," aired December 18, 2015. https://www.wnycstudios
.org/story/addiction

71 *85 percent of adults drink some kind of caffeinated beverage:* DC
Mitchell, CA Knight, J Hockenberry, et al., "Beverage Caffeine
Intakes in the U.S.," *Food and Chemical Toxicology* 63 (January
2014): 136–42.

5-Hour ENERGY shot or a look-alike: PJ Buckenmeyer, JA
Bauer, JF Hokanson, et al., "Cognitive Influence of a 5-h EN-
ERGY® Shot: Are Effects Perceived or Real?" *Physiology & Behav-
ior* 152, pt A (2015): 323–27.

performance was significantly improved: K Soar, E Chapman,
N Lavan, et al., "Investigating the Effects of Caffeine on Executive Functions Using Traditional Stroop and a New Ecologically-

Valid Virtual Reality Task, the Jansari Assessment of Executive Functions," *Appetite* 105 (2016): 156–63.

72 *In a study he led:* FG Vital-Lopez, S Ramakrishnan, TJ Doty, et al., "Caffeine Dosing Strategies to Optimize Alertness During Sleep Loss," *Journal of Sleep Research* 27, no. 5 (2018): e12711. doi: 10.1111/jsr.12711.

The State shall protect native and ancestral coca: J Briski, "Bolivia Says No to Cocaine, But Yes to Coca," *Christian Science Monitor,* March 20, 2012. https://www.csmonitor.com/World/Americas/Latin-America-Monitor/2012/0320/Bolivia-says-no-to-cocaine-but-yes-to-coca

73 *Sigmund Freud even wrote an 1884 treatise:* S Freud, "Über Coca," *Centralblatt für die ges. Therapie* 2 (1884): 289–314.

74 *substances not listed on the ingredients label:* J Tucker, T Fischer, L Upjohn, "Unapproved Pharmaceutical Ingredients Included in Dietary Supplements Associated with US Food and Drug Administration Warnings," *JAMA Network Open* 1, no. 6 (2018): e183337. doi:10.1001/jamanetworkopen.2018.3337

it reduced their fertility and sperm count: A Singh, SK Singh. "Evaluation of Antifertility Potential of Brahmi in Male Mouse," *Contraception* 79, no. 1 (2009): 71–79.

75 *the rate of crashes rose by 12 percent on April 20:* R Cohen, "Auto Crash Deaths Multiply After April 20 Cannabis Parties," *Reuters Health,* February 12, 2018. https://www.reuters.com/article/us-health-cannabis-traffic-safety/auto-crash-deaths-multiply-after-april-20-cannabis-parties-idUSKBN1FW2FV

76 *cannabis during adolescence:* MH Meier, A Caspi, A Ambler, et al., "Persistent Cannabis Users Show Neuropsychological Decline from Childhood to Midlife," *Proceedings of the National Academy of Sciences* 109, no. 40 (2012): E657–64.

the ones more likely to become stoners: NJ Jackson, JD Isen, R Khoddam, et al., "Impact of Adolescent Marijuana Use on Intelligence: Results from Two Longitudinal Twin Studies," *Proceedings of the National Academy of Sciences* 113, no. 5 (2016). doi:10.1073/pnas.1516648113

37 percent of 12th graders used it: National Institute on Drug Abuse, "Monitoring the Future Survey: High School and Youth Trends," December 2017. https://www.drugabuse.gov/publica tions/drugfacts/monitoring-future-survey-high-school-youth-trends

risk of becoming schizophrenic: RD Fields, "Link Between Adolescent Pot Smoking and Psychosis Strengthens," *Scientific American,* October 20, 2017. https://www.scientificamerican.com/article/link-between-adolescent-pot-smoking-and-psychosis-strengthens

A 2012 study of thirty nine sleep-deprived doctors: C Sugden, CR Housden, R Aggarwal, et al., "Effect of Pharmacological Enhancement on the Cognitive and Clinical Psychomotor Performance of Sleep-Deprived Doctors: A Randomized Controlled Trial," *Annals of Surgery* 255, no. 2 (2012): 222–27.

77 *modafinil's effects on healthy, non-sleep-deprived adults:* RM Battleday, AK Brem, "Modafanil for Cognitive Neuroenhancement in Healthy Non-Sleep-Deprived Subjects: A Systematic Review," *European Neuropsychopharmacology* 25, no. 11 (2015): 1865–81.

huge study conducted by Harold Kahn: HA Kahn, "The Dom Study of Smoking and Mortality Among US Veterans: Report on Eight and One-Half Years of Observations," *Epidemiological Approaches to the Study of Cancer and Other Chronic Diseases. Monograph No. 19.* National Cancer Institute (1996): 1–125.

78 *studies by neuroscientist Maryka Quik:* M Quik, T Bordia, D Zhang, et al., "Nicotine and Nicotinic Receptor Drugs: Potential for Parkinson's Disease and Drug-Induced Movement Disorders," *International Review of Neurobiology* 124 (2015): 247–71.

nicotine patch reduces impulsivity: AS Potter, PA News-house, "Acute Nicotine Improves Cognitive Deficits in Young Adults with Attention Deficit/Hyperactivity Disorder," *Pharmacology, Biochemistry and Behavior* 88, no. 4 (2008): 407–17.

told Neurology Today *in 2012:* D Hurley, "Growing List of Positive Effects of Nicotine Seen in Neurodegenerative Disorders," *Neurology Today* 12, no. 2 (2012): 37–38.

79 *6 percent of all high school seniors:* National Institute on Drug Abuse, "Misuse of Prescription Drugs." https://www.drugabuse.gov/publications/research-reports/misuse-prescription-drugs/what-scope-prescription-drug-misuse

20 percent of Ivy League students: D Kotz, "1 in 5 students at an Ivy League College Abuse Stimulant Drugs," *Boston Globe*, May 2, 2015. https://www.bostonglobe.com/lifestyle/health-wellness/2014/05/02/study-ivy-league-students-abuse-stimulant-drugs/vpaS16t8zh4pF8ga2zt69J/story.html

80 *they make people* feel *smarter:* I Ilieva, J Boland, MJ Farah, "Objective and Subjective Cognitive Enhancing Effects of Mixed Amphetamine Salts in Healthy People," *Neuropharmacology* 64 (2013): 496–505.

81 *One study from Stanford University:* JA Yesavage, MS Mumenthaler, JL Taylor, et al., "Donepezil and Flight Simulator Performance: Effects on Retention of Complex Skills," *Neurology* 59, no. 1 (2002): 123–25.

6. SLEEP ON IT

86 *Queen ants sleep about nine hours per night:* M Walker, "The Secrets of Ant Sleep Revealed," *BBC Earth News*, June 17, 2009. http://news.bbc.co.uk/earth/hi/earth_news/newsid_8100000/8100876.stm

Even jellyfish: RD Nath, CN Bedbrook, MJ Abrams, et al., "The

Jellyfish Cassiopea Exhibits a Sleep-Like State," *Current Biology* 27, no. 19 (2017): 2984–90.

87 *sleep can even increase problem-solving:* U Wagner, S Gais, H Haider, et al., "Sleep Inspires Insight," *Nature* 427 (2004): 352–55.
a recent, brilliant paper: GR Poe, "Sleep Is for Forgetting," *Journal of Neuroscience* 37, no. 3 (2017): 464–73.

88 *REM was discovered:* E Aserinsky, N Kleitman, "Regularly Occurring Periods of Eye Motility, and Concomitant Phenomena, During Sleep," *Science* 118, no. 3062 (1953): 273–74.
long, complex, and bizarre dreams persist: D. Oudiette, MJ Dealberto, G Uguccioni, et al., "Dreaming Without REM Sleep," *Consciousness and Cognition* 21, no. 3 (September 2012): 1129–40.
beginning with his 1899 book: S Freud, *The Interpretation of Dreams: The Complete and Definitive Text* (New York: Basic Books, 2010).

89 *Rechtschaffens devised an experiment:* A Rechtschaffens, BM Bergmann, "Sleep Deprivation in the Rat by the Disk-Over-Water Method," *Behavioral Brain Research* 69, no. 1–2 (1995): 55–63.

90 *analysis of sixteen prior studies:* FP Cappuccio, L D'Elia, P Strazzullo, et al., "Sleep Duration and All-Cause Mortality: A Systematic Review and Meta-Analysis of Prospective Studies," *Sleep* 33, no. 5 (2010): 585–92.
The Nurses' Health Study: NT Ayas, DP White, JE Manson, et al., "A Prospective Study of Sleep Duration and Coronary Heart Disease in Women," *Archives of Internal Medicine* 163, no. 2 (2003): 205–9.

91 *study from the American Heart Association:* J Fernandez-Mendoza, C LaGrotte, AN Vgontzas, et al., "Impact of Metabolic Syndrome on Mortality Is Modified by Objective Short Sleep Duration," *Journal of the American Heart Association* 6, no. 5 (May 17, 2017). pii: e005479. doi: 10.1161/JAHA.117.005479
recommendations from the National Sleep Foundation: National

Sleep Foundation, "National Sleep Foundation Recommends New Sleep Times." www.sleepfoundation.org/press-release/national-sleep-foundation-recommends-new-sleep-times/page/0/1

93 *able to peek inside our biological clock:* M Fox, "Body Clock Researchers Win Nobel Prize," *NBC News,* October 2, 2017. https://www.nbcnews.com/health/health-news/body-clock-researchers-win-nobel-prize-n806576
David E. Blask of the Laboratory of Chrono-Neuroendocrine Oncology: D Hurley, "For Our Own Good, Let There Be Dark," *Discover,* December 2016. http://discovermagazine.com/2016/dec/let-there-be-dark

94 *The best study I know of:* BL Uhlig, T Sand, SS Odegard, et al., "Prevalence and Associated Factors of DSM-V Insomnia in Norway: The Nord-Trøndelag Health Study (HUNT 3)," *Sleep Medicine* 15, no. 6 (June 2014): 708–13.

96 *Creative Dreaming:* P Garfield, *Creative Dreaming* (New York: Simon & Schuster, 1975).

98 *Even prescription drugs like Lunesta, Ambien, and Sonata:* "The Problem with Sleeping Pills," *Consumer Reports,* January 5, 2016. https://www.consumerreports.org/drugs/the-problem-with-sleeping-pills/
patients were evaluated before and after pineal gland removal: H Slawik, M Stoffel, L Riedl, et al., "Prospective Study on Salivary Evening Melatonin and Sleep Before and After Pinealectomy in Humans," *Journal of Biological Rhythms* 31, no. 1 (2016): 82–93.

7. JUST BREATHE

108 *the effects of mindful breathing on the brain:* HJ Scheibner, C Bogler, T Gleich, et al., "Internal and External Attention and the Default Mode Network," *Neuroimage* 148 (2017): 381–89.

109 *mindful breathing increased participants' white matter con-*

nections: YY Tang, Q Lu, X Geng, et al., "Short-Term Meditation Induces White Matter Changes in the Anterior Cingulate," *Proceedings of the National Academy of Science* 107, no. 35 (2010): 15649–52.

Breathing Above the Brain Stem: JL Herrero, S Khuvis, E Yeagle, et al., "Breathing Above the Brain Stem: Volitional Control and Attentional Modulation in Humans," *Journal of Neurophysiology* 119, no. 1 (2018): 145–59.

8. HOW TO HANDLE HEAD INJURIES

120 *That doesn't mean . . . that people are making it up:* D Hurley, "The Mystery Behind Neurological Symptoms Among US Diplomats in Cuba: Lots of Questions, Few Answers," *Neurology Today* 18, no. 6 (2018): 24–26.

121 *the center published a study in 2017:* J Mez, DH Daneshvar, PT Kiernan, et al., "Clinicopathological Evaluation of Chronic Traumatic Encephalopathy in Players of American Football," *Journal of the American Medical Association* 318, no. 4 (2017): 360–70.

Another study, by the U.S. Centers for Disease Control: National Institute for Occupational Safety and Health, "Heart Health Concerns for NFL Players," March 2012. https://www.cdc.gov/niosh/pgms/worknotify/pdfs/NFL_Notification_01-508.pdf

122 *Ewen's brain showed no physical signs of brain damage:* J Branch, "Autopsy Shows the N.H.L.'s Todd Ewen Did Not Have C.T.E.," *New York Times,* February 10, 2016. https://www.nytimes.com/2016/02/11/sports/hockey/autopsy-shows-the-nhls-todd-ewen-did-not-have-cte.html

123 *the definition of child abuse:* A Pawkowski, "'Concussion' Doctor Says Kids Shouldn't Play These Sports Until They're 18," *Today Show,* September 5, 2017. https://www.today.com/health/concussion-doctor-warns-against-contact-sports-kids-t115938

A study by researchers at McGill University: JS Delaney, VJ La-
croix, C Cagne, et al., "Concussions Among University Football
and Soccer Players: A Pilot Study," *Clinical Journal of Sport Medicine* 11, no. 4 (2001): 234–40.

125 *a study published in the journal* Pediatrics: DG Thomas, JN
Apps, RG Hoffmann, et al., "Benefits of Strict Rest After Acute
Concussion: A Randomized Controlled Trial," *Pediatrics* 135, no.
2 (2015): 213–23.

9. FOOD FOR THOUGHT

130 *The book he published in 1972:* RC Atkins, *Dr. Atkins' Diet Revolution: The High Calorie Way to Stay Thin Forever* (Philadelphia:
D. McKay Co., 1972).

131 *people who stuck to the MIND diet:* MC Morris, CC Tangney,
Y Wang, et al., "MIND Diet Associated with Reduced Incidence
of Alzheimer's Disease," *Alzheimer's & Dementia* 11, no. 9 (2015):
1007–14.

132 *a recent paper in* Nature Reviews Neuroscience: MP Mattson,
K Moehl, N Ghena, et al., "Intermittent Metabolic Switching,
Neuroplasticity and Brain Health," *Nature Reviews Neuroscience*
19, no. 2 (2018): 63–80.

135 *He then promptly died:* D Cavett, "When That Guy Died on My
Show," *New York Times,* May 3, 2007. https://opinionator.blogs
.nytimes.com/2007/05/03/when-that-guy-died-on-my-show/
Robert Atkins, he died at age 72: D Martin, "Robert C. Atkins,
72, Creator of Controversial Diet, Dies," *New York Times,* April 17,
2003. https://www.nytimes.com/2003/04/17/obituaries/robert-
c-atkins-72-creator-of-controversial-diet-dies.html
protect against the loss of memory: PM Herath, N Cherbuin, R
Eramudugolia, et al., "The Effect of Diabetes Medication on Cognitive Function: Evidence from the PATH Through Life Study,"
Biomed Research International (2016). doi: 10.1155/2016/7208429

10. HOW THE BRAIN HEALS ITSELF

142 *results of Schlaggar's experiment:* BL Schlaggar, DD O'Leary, "Potential of Visual Cortex to Develop an Array of Functional Units Unique to Somatosensory Cortex," *Science* 252, no. 5012 (1991): 1556–60.

148 *I came across a wild paper:* R Hamilton, JP Keenan, M Catala, et al., "Alexia for Braille Following Bilateral Occipital Stroke in an Early Blind Woman," *Neuroreport* 11, no. 2 (2000): 237–40.

11. THE BIONIC BRAIN

155 *Delgado said he had proved:* JA Osmundsen, "'Matador' with a Radio Stops Wired Bull; Modified Behavior in Animals Subject of Brain Study," *New York Times,* May 17, 1965: 1.

He eventually published his views in a book: JMR Delgado, *Physical Control of the Mind: Toward a Psychocivilized Society* (New York: Harper & Row, 1969).

161 *the study in* Annals of Neurology: C Hamani, MP McAndrews, M Cohn, et al., "Memory Enhancement Induced by Hypothalamic/Fornix Deep Brain Stimulation," *Annals of Neurology* 63, no. 1 (2008): 119–23.

a team of researchers at UCLA and Tel Aviv University in Israel tested DBS: N Suthana, Z Haneef, J Stern, et al., "Memory Enhancement and Deep-Brain Stimulation of the Entorhinal Area," *New England Journal of Medicine* 366, no. 6 (2012): 502–10.

But a larger study: J Jacobs, J Miller, SA Lee, et al., "Direct Electrical Stimulation of the Human Entorhinal Region and Hippocampus Impairs Memory," *Neuron* 92, no. 5 (2016): 983–90.

163 *BrainGate, one of the most promising devices:* S Boseley, "Paralyzed Man Moves Arm Using Power of Thought in World First," *The Guardian,* March 29, 2017. https://www.theguardian.com/science/2017/mar/28/neuroprosthetic-tetraplegic-man-control-hand-with-thought-bill-kochevar

volunteer named Nathan Copeland: SN Flesher, JL Collinger,

ST Foldes, et al., "Intracortical Microstimulation of Human Somatosensory Cortex," *Science Translational Medicine* 8, no. 361 (October 9, 2016): 361ra141.

165 *The company states:* https://www.neuralink.com/
Tim Urban spoke with Musk: T Urban, "Neuralink and the Brain's Magical Future," *Wait but Why,* April 20, 2017. https://waitbutwhy.com/2017/04/neuralink.html
closest to mental telepathy yet achieved: C Grau, R Ginhoux, A Rivera, et al., "Conscious Brain-to-Brain Communication in Humans Using Non-Invasive Technologies," *PLOS ONE* 9, no. 8 (2014). doi: 10.1371/journal.pone.0105225

12. SHOCK AND TINGLE

169 *One of the largest and most recent:* CK Loo, MM Husain, WM McDonald, et al., "International Randomized-Controlled Trial of Transcranial Direct Current Stimulation in Depression," *Brain Stimulation* 11, no. 1 (2018): 125–33.

170 *a type of tDCS improved the math skills:* A Snowball, I Tachtsidis, T Popescu, et al., "Long-Term Enhancement of Brain Function and Cognition Using Cognitive Training and Brain Stimulation," *Current Biology* 23, no. 11 (2013): 987–92.

171 *find relief in days or weeks:* Johns Hopkins Medicine, "Frequently Asked Questions about ECT." https://www.hopkinsmedicine.org/psychiatry/specialty_areas/brain_stimulation/ect/faq_ect.html

172 *Benjamin Franklin claimed:* S Finger, "Benjamin Franklin and the Neurosciences," Ottorino Rossi Award lecture (2006). https://www.functionalneurology.com/materiale_cic/131_XXI_2/1177_Benjamin/index.html
insulin was given in large doses to people with schizophrenia: K Jones, "Insulin Coma Therapy in Schizophrenia," *Journal of the Royal Society of Medicine* 93 (2000): 147–49.
an Italian psychiatrist, Ugo Cerletti: U Cerletti, L Bini,

"L'Elettroshock," *Rivista Sperimentale di Frenatria* 1 (1940): 209–310.

175 *study by researchers at Rush University:* KA Skarupski, CC Tangney, H Li, et al., "Mediterranean Diet and Depressive Symptoms Among Older Adults Over Time," *Journal of Nutrition, Health & Aging* 17, no. 5 (2013): 441–45.

when depressed people are counseled by nutritionists: FN Jacka, A O'Neill, R Opie, et al., "A Randomized Controlled Trial of Dietary Improvement for Adults with Major Depression (the 'SMILES' Trial)," *BMC Medicine* 15, no. 1 (2017): 23. doi: 10.1186/s12916-017-0791-y

13. STEM CELLS AND BEYOND

177 *first-in-human clinical trial:* J Portnow, TW Synold, B Badie, et al., "Neural Stem Cell–Based Anticancer Gene Therapy: A First-in-Human Study in Recurrent High-Grade Glioma Patients," *Clinical Cancer Research* 23, no. 12 (2017): 2951–60.

184 *Scientists were recently blown away by a study:* M Ortiz-Virumbrales, CL Moreno, I Kruvlikov, et al., "CRISPR/Cas9-Correctable Mutation-Related Molecular and Physiological Phenotypes in iPSC-Derived Alzheimer's PSEN2N141I Neurons," *Acta Neuropathologica Communcations* 5 (2017): 77. doi: 10.1186/s40478-017-0475-z

182 *Gage published a revolutionary paper:* PS Eriksson, E Perfilieva, T Bjork-Eriksson, et al., "Neurogenesis in the Adult Human Hippocampus," *Nature Medicine* 4, no. 11 (1998): 1313–17.

183 *no signs of new brain cells after all:* SF Sorrells, MF Paredes, A Cebrian-Silla, et al., "Human Hippocampal Neurogenesis Drops Sharply in Children to Undetectable Levels in Adults," *Nature* 15, no. 555 (2018): 377–81.

187 *made blind mice see:* JH Lim, BK Stafford, PL Nguyen, et al., "Neural Activity Promotes Long-Distance, Target-Specific Re-

generation of Adult Retinal Axons," *Nature Neuroscience* 19, no. 8 (2016): 1073–84.

14. THE YOUNGER BRAIN

199 *paper by researchers at the University of Texas and UC San Diego:* AF Ward, K Duke, A Gneezy, et al., "Brain Drain: The Mere Presence of One's Own Smartphone Reduces Available Cognitive Capacity," *Journal of the Association for Consumer Research* 2, no. 2 (2017). https://www.journals.uchicago.edu/doi/abs/10.1086/691462

203 *adolescents who don't get enough sleep:* EM Cespedes Feliciano, M Quante, SL Rifas-Shiman, et al., "Objective Sleep Characteristics and Cardiometabolic Health in Young Adolescents," *Pediatrics* 142, no. 1 (2018). doi: 10.1542/peds.2017-4085
American Academy of Pediatrics recommends: "American Academy of Pediatrics Announces New Safe Sleep Recommendations to Protect Against SIDS, Sleep-Related Infant Deaths," October 24, 2016. https://www.aap.org/en-us/about-the-aap/aap-press-room/pages/american-academy-of-pediatrics-announces-new-safe-sleep-recommendations-to-protect-against-sids.aspx

206 *pediatricians recently wrote:* ER Cheng, LG Fiechtner, AE Carroll, "Seriously, Juice Is Not Healthy," *New York Times,* July 7, 2018. https://www.nytimes.com/2018/07/07/opinion/sunday/juice-is-not-healthy-sugar.html

15. THE OLDER BRAIN

216 *the neurobiology of wisdom:* TW Meeks, DV Jeste, "Neurobiology of Wisdom: A Literature Review," *Archives of General Psychiatry* 66, no. 4 (2009): 355–65. J Reichstadt, G Sengupta, CA Depp, et al., "Older Adults' Perspectives on Successful Aging: Qualitative Interviews," *American Journal of Geriatric Psychiatry* 18, no. 7 (2010): 567–75.

216 *greater in those with online social networks:* R Kanai, B Bahrami, R Roylance, et al., "Online Social Network Size Is Reflected in Human Brain Structure," *Proceedings of the Royal Society B: Biological Sciences* 279, no. 1732 (2012): 1327–34.

217 *12 percent reduced risk of Alzheimer's:* JM Zissimopoulos, D Barthold, RD Brinton, et al., "Sex and Race Differences in the Association Between Statin Use and the Incidence of Alzheimer Disease," *JAMA Neurology* 74, no. 2 (2017): 225–32.

218 *recent study of nearly 13,000 older adults:* AM Rawlings, AR Sharrett, TH Mosley, et al., "Glucose Peaks and the Risk of Dementia and 20-Year Cognitive Decline," *Diabetes Care* 40, no. 7 (2017): 879–86.

219 *Super Agers study:* L Neergaard, "Superagers' Brains Offer Clues for Sharp Memory in Old Age," *US News & World Report,* February 22, 2018. https://www.usnews.com/news/best-states/california/articles/2018-02-22/superagers-brains-offer-clues-for-sharp-memory-in-old-age

220 *loneliness... puts people at higher risk of cognitive loss:* NJ Donovan, Q Wu, DM Rentz, et al., "Loneliness, Depression and Cognitive Function in Older U.S. Adults," *International Journal of Geriatric Psychiatry* 32, no. 5 (2017): 564–73.

resistance training with weights improves cognition: JR Best, BK Chiu, C Liang Hsu, et al., "Long-Term Effects of Resistance Exercise Training on Cognition and Brain Volume in Older Women: Results from a Randomized Controlled Trial," *Journal of the International Neuropsychological Society* 21, no. 10 (2015): 745–56.

benefits on cognition were seen in two clinical trials: JA Mortimer, D Ding, AR Bornstein, et al., "Changes in Brain Volume and Cognition in a Randomized Trial of Exercise and Social Interaction in a Community-Based Sample of Non-Demented Chinese

Elders," *Journal of Alzheimers Disease* 30, no. 4 (2012): 757–66.
ST Cheng, PK Chow, YQ Song, et al., "Mental and Physical Activities Delay Cognitive Decline in Older Persons with Dementia," *American Journal of Geriatric Psychiatry* 22, no. 1 (2014): 63–74.

INDEX